Mathematical Analysis of Problems in the Natural Sciences

Vladimir Zorich

Mathematical Analysis of Problems in the Natural Sciences

Translated by Gerald Gould

Vladimir Zorich
Department of Mathematics
(Mech-Math)
Moscow State University
Vorobievy Gory
119992 Moscow
Russia
vzor@mccme.ru

Translator
Gerald G. Gould
School of Mathematics
Cardiff University
Senghenydd Road
CF24 4AG Cardiff
UK

Original Russian edition *Matematicheskij analiz zadach estestvoznaniya*
published by MCCME, Moscow, Russia, 2008

ISBN 978-3-642-14812-5 e-ISBN 978-3-642-14813-2
DOI 10.1007/978-3-642-14813-2
Springer Heidelberg Dordrecht London New York

Library of Congress Control Number: 2010938125

Mathematics Subject Classification (2010): 00A69, 00A73, 51Pxx, 76Fxx, 80Axx, 94-xx

Cover design: deblik

Printed on acid-free paper

Springer is part of Springer Science+Business Media (www.springer.com)

Contents

Part III Classical Thermodynamics and Contact Geometry

Synopsis

This, in a shortened form, is a write-up of an annual special course on natural sciences. In it three themes (called parts) are presented:

The analysis of dimensions of physical quantities with examples of applications including Kolmogorov's model for turbulence.

Functions of a very large number of variables and the principle of concentration: the non-linear law of large numbers, the geometric meaning of the Gauss and Maxwell distributions and the Kotel'nikov–Shannon theorem.

Classical thermodynamics and contact geometry: two principles of thermodynamics in the language of differential forms, contact distributions and the Frobenius theorem, and the Carnot–Carathéodory metric.

This special course is in the first instance intended for mathematicians, but may also be useful to students and specialists of other disciplines.

Located in the Appendix is the author's popular article "Mathematics as language and method".

Preface

This is a short course of natural-sciences content intended for mathematicians. It has a potential for further development in different directions.

Here I present something from the legacy of Galileo, Newton, Euler, Bernoulli, Carnot, Clausius, Boltzmann, Gibbs, Poincaré, Einstein, Planck, Schrödinger, Carathéodory, Kolmogorov, Kotel'nikov, Shannon, and others.

Certainly, the title "Mathematical Analysis of problems in the Natural Sciences"reflects only a trend rather than a promise of any universality, like "everything, at once, and for free". The selection of the three subjects for the book is very conditional.

Note that those who are held in particularly high esteem by us, mathematicians, such as Archimedes, Newton, Leibniz, Euler, Gauss, Poincaré, were not mere mathematicians, but also scientists, natural philosophers.

In mathematics, solving concrete problems and creating abstract general theories are inseparable processes, like inhaling and exhaling. A prolonged violation of a balance between them is extremely dangerous. One should avoid getting into a position of a fisherman who enthusiastically pulls the fishing line and continues fishing on an ice floe that is already drifting away from the shore.

As Hermann Weyl noted, "a truly realistic mathematics should be, in conjunction with physics, a part of theoretical construction of a unified real world". By the way, this unity of physics and mathematics is still reflected in Russian diplomas of Candidates[1]and Doctors of Mathematical and Physical Sciences.

In conclusion I should like to thank all those who have helped with the proof reading of the original text, especially V.I. Arnold, who did not skip a single paragraph of the hundred-page print-out[2] and made a large number of incisive observations accompanied by comments.[3] If I have not taken into account all remarks or wishes of colleagues here, then this does not mean that I have ignored them; rather, I regard them as a subject of further deliberations and discussions.

V. Zorich

---

[1] *Translator's note*: The Kandidat degree is roughly equivalent to a Ph.D. and the Doktor degree is roughly equivalent to a D.Sc.

[2] This is no exaggeration: the file intended for V.I. Arnold was accidentally sent to the printer which systematically replaced some letters from Cyrillic to something different (sometimes with quite amusing results). And Arnold also corrected all these printing errors.

[3] For example, that the arguments in Part I are 'obscurantist'. With regard to this I should alert the reader. See also the footnote on p. 19.

Part I
Analysis of Dimensions of Physical Quantities

A few introductory words

An abstract number, for example, 1 or $2\frac{2}{3}$, and the arithmetic of abstract numbers, for example, that $2 + 3 = 5$ irrespective of whether one is adding apples or elephants, is a great achievement of civilization comparable with the invention of writing. We have become so used to this that we are no longer aware of the miracle lying somewhere at the very foundation of the amazing effectiveness of mathematics.

If you know what object a number refers to, then, as a rule, at the same time there immediately arise additional possibilities and constraints. We recall the nursery rhyme: "And I have obtained in my answer two navvies and two thirds". Yes, the number $2\frac{2}{3}$ is permissible in arithmetic but is not permissible in this concrete situation.

How does one make use of the fact that in concrete problems we do not deal with abstract numbers but rather with dimensional quantities ?

Is there a science taking account of this ? Not a great one, but yes. Every qualified natural scientist knows it (as well as the dangers of the unskillful use of it). And this is what we shall be talking about.[4]

[4] For the information of the reader: the numbering of the formulae is continuous within each chapter, but starts afresh in subsequent chapters.

Chapter 1
Elements of the theory

1.1 Dimension of a physical quantity (preliminaries)

1.1.1 Measurement, unit of measurement, measuring process

All the above concepts are fundamental and have been subjected to analysis by the best representatives of science, primarily physicists and mathematicians. This entails the analysis of the concepts of space, solid body, motion, time, causality, and so on.

At this stage we do not intend to get too deeply involved in this; however, we note that each theory provides a good model of some sphere of phenomena only in certain scales. Unfortunately, sometimes we only know what these scales are when (and only when) the theory ceases to conform to reality. At those moments we usually return to the fundamentals of the theory again, subjecting it to a thorough analysis and appropriate reconstruction.

So let us now start with the accumulation of the useful concrete material available to us.

1.1.2 Basic and derived units

In life we constantly use certain units of measurement of length, mass, time, velocity, energy, power, and so on. We single out some of them as basic units, while others turn out to be derived units.

Examples of basic units are the unit L of length, which is the *metre* (denoted by m); the unit M of mass, which is the *kilogramme* (denoted by kg) and the unit T of time, which is the *second* (denoted by s).

Examples of derived units are v velocity (m/s) $[v] = LT^{-1}$; V volume (m^3) $[V] = L^3$; a acceleration (m/s^2) $[a] = LT^{-2}$; l light year $[l] = [cT] = L$; F force, $F = Ma$, $[F] = [Ma] = MLT^{-2}$.

V. Zorich, *Mathematical Analysis of Problems in the Natural Sciences*,
DOI 10.1007/978-3-642-14813-2_1, © Springer-Verlag Berlin Heidelberg 2011

In all these examples we observe the formula $L^{d_1}M^{d_2}T^{d_3}$ for the dimension of a mechanical physical quantity; $\{d_1, d_2, d_3\}$ is the dimension vector in the basis $\{L, M, T\}$.

We shall develop this vectorial algebro-geometric analogy.

1.1.3 *Dependent and independent units*

Example. The units of the quantities v, a and F are independent and can also be taken as the basic units; this is because $[L] = v^2a^{-1}$, $[M] = Fa^{-1}$, $[T] = va^{-1}$.

We may guess analogies with vector space, its basis and systems of independent vectors. (A deeper sense of this analogy will be revealed later.)

1.2 A formula for the dimension of a physical quantity

1.2.1 *Change of the numerical values of a physical quantity under a change of the sizes of the basic units.*

Example. If distance is measured in kilometres (that is, instead of 1 metre as the unit of measurement of lengths one takes 1 kilometre as the unit, which is 1000 times greater), then the same physical length will have two different numerical values with respect to these two units of length, namely, $1\text{km} = 10^3\text{m}$, $L\text{km} = 10^3L\text{m}$, $1\text{m} = 10^{-3}\text{km}$, $L\text{m} = 10^{-3}L\text{km}$. Thus, a change of the unit of length by α times leads to a change in the numerical value of Lm of all lengths being measured by α^{-1} times, that is, the value L is replaced by $\alpha^{-1}L$.

This also relates to possible changes of the units of mass and time (tonne, gramme, milligramme; hour, day, year, millisecond, and so on).

Hence, if a physical quantity has dimensions $\{L^{d_1}M^{d_2}T^{d_3}\}$ in the basis $\{L, M, T\}$, that is, $\{d_1, d_2, d_3\}$ is its dimension vector, then the change in the units of measurement of length, mass and time by α_1, α_2, α_3 times, respectively, must entail a change in the numerical value of this quantity by $\alpha_1^{-d_1}\alpha_2^{-d_2}\alpha_3^{-d_3}$ times.

1.2.2 *Postulate of the invariance of the ratio of the values of physical quantities with the same name.*

Example. The area of a triangle is a function $y = f(x_1, x_2, x_3)$ of the lengths of its three sides. We take another triangle and calculate its area $\tilde{y} = f(\tilde{x}_1, \tilde{x}_2, \tilde{x}_3)$.

Under a change of the size of the unit of length the numerical values of y and $\tilde{y}$ change. However, their ratio $y/\tilde{y}$ remains unaltered.

Suppose now that we have two physical quantities $y = f(x_1, x_2, \ldots, x_m)$, $\tilde{y} = f(\tilde{x}_1, \tilde{x}_2, \ldots, \tilde{x}_m)$, depending not merely on three quantities, but now on a finite collection of quantities involving just length, mass or time. Then we get the following fundamental postulate of dimension theory.

POSTULATE OF THE ABSOLUTENESS OF RATIOS

$$\frac{\tilde{y}}{y} = \frac{f(\tilde{x}_1, \tilde{x}_2, ..., \tilde{x}_m)}{f(x_1, x_2, ..., x_m)} = \frac{f(\alpha\tilde{x}_1, \alpha\tilde{x}_2, ..., \alpha\tilde{x}_m)}{f(\alpha x_1, \alpha x_2, ..., \alpha x_m)}. \tag{1.1}$$

In other words, it is postulated that under a change of scale of a basic unit (length, mass, time, ...) physical quantities y, $\tilde{y}$ of the same type (all areas, all volumes, all velocities, all forces, ...) change their numerical values in the same proportion (its own for each type of quantity).

1.2.3 *Function of dimension and a formula for the dimension of a physical quantity in a given basis.*

Equation 1.1 shows that the ratio

$$\frac{f(\alpha x_1, \alpha x_2, ..., \alpha x_m)}{f(x_1, x_2, ..., x_m)} =: \varphi(\alpha)$$

depends only on α. This enables one to indicate the form of the function $\varphi(\alpha)$. First we note that

$$\frac{\varphi(\alpha_1)}{\varphi(\alpha_2)} = \varphi\left(\frac{\alpha_1}{\alpha_2}\right). \tag{1.2}$$

Indeed,

$$\frac{\varphi(\alpha_1)}{\varphi(\alpha_2)} = \frac{f(\alpha_1 x_1, \alpha_1 x_2, ..., \alpha_1 x_m)}{f(\alpha_2 x_1, \alpha_2 x_2, ..., \alpha_2 x_m)} = \frac{f(\frac{\alpha_1}{\alpha_2} x_1, \frac{\alpha_1}{\alpha_2} x_2, ..., \frac{\alpha_1}{\alpha_2} x_m)}{f(x_1, x_2, ..., x_m)} = \varphi\left(\frac{\alpha_1}{\alpha_2}\right).$$

Here the left and right equalities follow immediately from the definition of the function φ; and, taking into account the independence of φ of the variables $(x_1, x_2, ..., x_m)$, one can go over to the variables $\alpha_2^{-1}(x_1, x_2, ..., x_m)$ and obtain the inner equality.

Assuming that φ is a regular function we differentiate (1.2) with respect to α_1 and then setting $\alpha_1 = \alpha_2$ we obtain the equation

$$\frac{1}{\varphi(\alpha)}\frac{d\varphi}{d\alpha} = \frac{1}{\alpha}\varphi'(1).$$

Since $\varphi(1) = 1$, the solution of this equation with this initial condition has the form

$$\varphi(\alpha) = \alpha^d.$$

Thus, the above form of the function φ is a consequence of our postulate of the theory of dimensions of physical quantities. In our formulation of the postulate we have supposed (to keep our initial considerations simple) that all the variables $(x_1, x_2, ..., x_m)$ have the same character (length or mass or time or velocity, ...). But, clearly, any functional dependence of a physical quantity on a collection of physical variables can be represented in the form

$$y = f(x_1, x_2, ..., x_{m_1}, y_1, y_2, ..., y_{m_2}, ..., z_1, z_2, ..., z_{m_k}),$$

where physical variables of the same type are collected in a group and denoted by a common symbol (we did not want to introduce a numbering of the groups). This means that

$$\frac{f(\alpha_1 x_1, \alpha_1 x_2, ..., \alpha_1 x_{m_1}, ..., \alpha_k z_1, \alpha_k z_2, ..., \alpha_k z_{m_k})}{f(x_1, x_2, ..., x_{m_1}, ..., z_1, z_2, ..., z_{m_k})} = \alpha_1{}^{d_1} \cdot ... \cdot \alpha_k{}^{d_k},$$

if the variables in different groups are allowed to have independent changes of scale. In this case, within the scope of the postulate we have found the law

$$\begin{aligned} &f(\alpha_1 x_1, \alpha_1 x_2, ..., \alpha_1 x_{m_1}, ..., \alpha_k z_1, \alpha_k z_2, ..., \alpha_k z_{m_k}) = \\ &= \alpha_1{}^{d_1} \cdot ... \cdot \alpha_k{}^{d_k}\, f(x_1, x_2, ..., x_{m_1}, ..., z_1, z_2, ..., z_{m_k}) \end{aligned} \tag{1.3}$$

for the change of the numerical value of the physical quantity

$$y = f(x_1, x_2, ..., x_{m_1}, ..., z_1, z_2, ..., z_{m_k})$$

under a change of the scales of the units of physically independent variables.

The tuple $(d_1, \ldots, d_k)$ is called the *dimension vector* or simply the *dimension* of the quantity y with respect to the distinguished independent physical units of measurement. The function $\varphi(\alpha_1, ..., \alpha_k) = \alpha_1{}^{d_1} \cdot ... \cdot \alpha_k{}^{d_k}$ is called the *dimension function*.

The symbol $[y]$ denotes either the dimension vector or the dimension function depending on the context. A physical quantity is called *dimensionless* if its dimension vector is zero. For example, if $\varphi(\alpha_1, ..., \alpha_k) = \alpha_1{}^{d_1} \cdot ... \cdot \alpha_k{}^{d_k}$ is the dimension function of a physical quantity y with respect to independent physical quantities $\{x_1, \ldots, x_k\}$, then the ratio

$$\Pi := \frac{y}{{x_1}^{d_1} \cdot \ldots \cdot {x_k}^{d_k}}$$

is a dimensionless quantity with respect to $\{x_1, \ldots, x_k\}$.

1.3 Fundamental theorem of dimension theory

1.3.1 The Π-Theorem.

We now consider the general case of the dependence

$$y = f(x_1, x_2, \ldots, x_k, \ldots, x_n) \tag{1.4}$$

of the physical quantity y on the variable quantities $x_1, x_2, \ldots, x_n$, among which only the first k are physically (dimensionally) independent in the sense of our theory of the dimensions of physical quantities.

By taking $x_1, x_2, \ldots, x_k$ as the units of measurement of the corresponding quantities, that is, changing the scale by setting $\alpha_1 = {x_1}^{-1}, \ldots, \alpha_k = {x_k}^{-1}$, equation (1.3) yields the relation

$$\Pi = f(1, \ldots 1, \Pi_1, \Pi_2, \ldots, \Pi_{n-k}) \tag{1.5}$$

between the dimensionless quantities

$$\Pi = \frac{y}{{x_1}^{d_1} \cdot \ldots \cdot {x_k}^{d_k}}, \quad \Pi_i = \frac{x_{k+i}}{{x_1}^{d_{i_1}} \cdot \ldots \cdot {x_k}^{d_{i_k}}},$$

where $i = 1, \ldots, n-k$. Equation (1.5) can be rewritten in the form

$$y = {x_1}^{d_1} \cdot \ldots \cdot {x_k}^{d_k} \cdot f(1, \ldots, 1, \Pi_1, \Pi_2, \ldots, \Pi_{n-k}). \tag{1.6}$$

Thus, by using the scale homogeneity of the dependences between physical quantities expressed in the postulate formulated above we can go over from relation (1.4) to the relation (1.5) between dimensionless quantities and, in so doing, reduce the number of variables; or we can go over to the relation (1.6), which is equivalent to relation (1.5), by explicitly selecting the entire dimensional constituent of y with respect to the maximal system $x_1, x_2, \ldots, x_k$ of dimensionally independent variables.

The possibility of such a transition from the general relation (1.4) to the simpler relations (1.5) and (1.6) forms the content of the so-called Π-*Theorem*,[1] which is the fundamental theorem of the theory of dimensions of physical quantities (which we have just proved).

1.3.2 Principle of similarity.

The content, sense and capability of the Π-Theorem, as well as the pitfalls associated with it, will be developed below by way of concrete examples of its use. However, one idea of the effective (and striking) application of the above theorem lies right on the surface and is obvious: without damaging aeroplanes, ships and other such objects, many experiments can be carried out in the laboratory on models, after which the results (for example, the experimentally found dimensionless dependence (1.5)) can be recomputed with the aid of the Π-Theorem for the actual objects of natural size (via formula (1.6)).

[1] It is also called Buckingham's Theorem, in connection with the appearance of the papers by E. Buckingham "On physically similar systems; illustrations of the use of dimensional equations", Phys. Rev. **4** (1914), 345–376, and E. Buckingham "The principle of similitude", Nature **96** (1915), 396–397. In implicit form the Π-Theorem and the Similarity Principle are also contained in the paper J. H. Jeans, Proc. Roy. Soc. **76** (1905), 545. In a sense the laws of similarity were essentially known to Newton and Galileo. A history of this question can be found in the paper [5] along with a reference to the paper [6], where there is everything except the name Π-*Theorem*. In this connection we also mention a slightly curious publication [7].

Chapter 2
Examples of applications

We now consider some examples in which various aspects of the Π-Theorem are explained.

2.1 Period of rotation of a body in a circular orbit (laws of similarity)

A body of mass m is kept in a circular orbit of radius r by a central force F. It is required to find the period of rotation

$$P = f(r, m, F).$$

Here and in what follows we fix the basis of fundamental physical units that is standard in mechanics, namely, length, mass and time, which, following Maxwell, we denote by $\{L, M, T\}$. (In thermodynamics the symbol T is used to denote absolute temperature, but unless otherwise stated, we shall meanwhile use this notation for the unit of time.)

Let us find the dimension vector of the quantities P, r, m, F in the basis $\{L, M, T\}$. We write them as the columns of the following table:

	P	r	m	F
L	0	1	0	1
M	0	0	1	1
T	1	0	0	-2

Since, as we have shown, the dimension function always has degree form, multiplication of such functions corresponds to addition of the degree exponents, in other words, they correspond to linear operations on the dimension vectors of the corresponding physical quantities.

DOI 10.1007/978-3-642-14813-2_2,

Hence, using standard linear algebra, one can find a system of independent quantities from the matrix formed by their dimension vectors; also, by decomposing the dimension vector of some quantity into the dimension vectors of the selected independent quantities, one can find a formula for the dimension of this quantity in the system of independent quantities of the concrete problem.

Thus, in our case the quantities r, m, F, are independent because the matrix formed by the vectors $[r]$, $[m]$, $[F]$ is non-singular. Finding the expansion $[P] = \frac{1}{2}[r] + \frac{1}{2}[m,] - \frac{1}{2}[F]$ on the basis of formula (1.6) of Chapter I we immediately see that

$$P = \left(\frac{mr}{F}\right)^{1/2} \cdot f(1,1,1).$$

Thus, to within the positive factor $c = f(1,1,1)$ (which can be found by a single laboratory experiment) we have found the dependence of P on r, m, F. Of course, knowing Newton's law $F = m \cdot a$, in the present instance we could easily have found the final formula, where $c = 2\pi$. However, everything that we have used is just a general indication of the existence of the dependence $P = f(r, m, F)$.

2.2 The gravitional constant. Kepler's third law and the degree exponent in Newton's law of universal gravitation.

After Newton we find the degree exponent α in the law of universal gravitation

$$F = G\frac{m_1 m_2}{r^\alpha}.$$

We use the previous problem and Kepler's third law (which was known to Newton) which, for circular orbits, implies that the square of the periods of rotation of the planets (with respect to a central body of mass M) are proportional to the cubes of the radii of their orbits. In view of the result of the previous problem and the law of universal gravitation (with the exponent α not yet found) we have

$$\left(\frac{P_1}{P_2}\right)^2 = \left(\frac{m_1 r_1}{F_1}\right)^{1/2} / \left(\frac{m_2 r_2}{F_2}\right)^{1/2} = \left(\frac{m_1 r_1}{\frac{m_1 M}{r^\alpha}}\right)^{1/2} / \left(\frac{m_2 r_2}{\frac{m_2 M}{r^\alpha}}\right)^{1/2} = \left(\frac{r_1}{r_2}\right)^{\alpha+1}.$$

But by Kepler's law $\left(\frac{P_1}{P_2}\right)^2 = \left(\frac{r_1}{r_2}\right)^3$. Hence $\alpha = 2$.

2.3 Period of oscillation of a heavy pendulum (inclusion of g).

After the detailed explanations involved in the solution of the first problem we can now allow ourselves a more compact account, pausing only at certain new circumstances.

We shall find the period of oscillation of a pendulum. More precisely, a load of mass m is fixed at the end of a weightless suspension of length l inclined from the equilibrium position at some initial angle φ_0, is let go and under the action of the force of gravity starts to perform a periodic oscillation. We shall find the period P of these oscillations.

To write $P = f(l, m, \varphi_0)$ would be wrong because a pendulum has different periods of oscillation on the Earth and the Moon in view of the difference in the forces of gravity at the surfaces of these two bodies. The force of gravity at the surface of a body, for example, the Earth, is characterized by the quantity g which is the acceleration of free fall at the surface of this body. Therefore, instead of the impossible relation $P = f(l, m, \varphi_0)$, one must assume that $P = f(l, m, g, \varphi_0)$.

We write the dimension vectors of all these quantities in the basis $\{L, M, T\}$:

	P	l	m	g	φ_0
L	0	1	0	1	0
M	0	0	1	0	0
T	1	0	0	−2	0

Clearly the vectors $[l]$, $[m]$, $[g]$ are independent and $[P] = \frac{1}{2}[l] - \frac{1}{2}[g]$.

In view of the Π-Theorem in the form of relation 1.6 of Chapter I, it follows that

$$P = \left(\frac{l}{g}\right)^{\frac{1}{2}} \cdot f(1, 1, 1, \varphi_0).$$

We have found that $P = c(\varphi_0) \cdot \sqrt{\frac{l}{g}}$, where the dimensionless factor $c(\varphi_0)$ depends only on the dimensionless angle φ_0 of the initial inclination (measured in radians).

The precise value of $c(\varphi_0)$ can also be found, although this time this is no longer all that easy. It can be done by solving the equation of motion of an oscillating heavy pendulum and invoking the elliptic integral

$$F(k, \varphi) := \int_0^{\varphi} \frac{d\theta}{\sqrt{1 - k^2 \sin^2 \theta}}.$$

Namely, $c(\varphi_0) = 4K(\sin(\frac{1}{2}\varphi_0))$, where $K(k) := F(k, \frac{1}{2}\pi)$.

2.4 Outflow of volume and mass in a waterfall

On a broad shelf having the form of a step on the upper platform, water falling under the action of gravity forms a waterfall. The depth of the water on the upper platform is known and is equal to h. It is required to find the specific volumetric outflow V (per unit of time on a unit of width of the step) of the water. If we look at the mechanism of the phenomenon in the right way, then we see that $V = f(g,h)$.

Since this phenomenon is determined by gravitation, along with the dimensional constant g (free-fall acceleration), we could, as a precaution, introduce the density ϱ of the fluid, that is, we suppose that

$$V = f(\varrho,g,h).$$

We now carry out the standard procedure of finding the dimension vectors:

	V	ϱ	g	h
L	2	−3	1	1
M	0	1	0	0
T	−1	0	−2	0

Clearly the vectors $[\varrho],[g],[h]$ are independent and $[V] = \frac{1}{2}[g] + \frac{3}{2}[h]$.

In view of the Π-Theorem we now obtain that

$$V = g^{\frac{1}{2}}h^{\frac{3}{2}} \cdot f(1,1,1).$$

Thus, $V = c \cdot g^{\frac{1}{2}}h^{\frac{3}{2}}$, where c is a constant to be determined, for example, in a laboratory experiment. Here the specific outflow Q of the mass is clearly equal to ϱV. One could also have arrived at the same formula by applying the method of dimensions to the relation $Q = f(\varrho,g,h)$.

2.5 Drag force for the motion of a ball in a non-viscous medium

A ball of radius r moves with velocity v in a non-viscous medium of density ϱ. It is required to find the drag force acting on the ball. (One could, of course, assume that there is a flow moving with velocity v past a ball at rest, which is a typical situation in wind-tunnel tests.)

We write down the general formula $F = f(\varrho,v,r)$ and analyse it in terms of dimensions:

	F	ϱ	v	r
L	1	−3	1	1
M	1	1	0	0
T	−2	0	−1	0

Clearly the vectors $[\varrho], [v], [r]$ are independent and $[F] = [\varrho] + 2[v] + 2[r]$. In view of the Π-Theorem we now obtain

$$F = \varrho v^2 r^2 \cdot f(1,1,1). \tag{2.1}$$

Thus, $F = c \cdot \varrho v^2 r^2$, where c is a dimensionless constant coefficient.

2.6 Drag force for the motion of a ball in a viscous medium

Before we turn to the formulation of this problem we recall the notion of viscosity of a medium and find the dimension of viscosity.

If one places a sheet of paper on the surface of thick honey, then in order to move the sheet along the surface one needs to apply certain forces. In first approximation the force F applied to the sheet stuck on the surface of the honey will be proportional to the area S of the sheet, the speed v of its motion and inversely proportional to the distance h from the surface to the bottom where the honey is also stuck and stays motionless in spite of the motion at the top (like a river).

Thus, $F = \eta \cdot Sv/h$. The coefficient η in this formula depends on the medium (honey, water, air, and so on) and is called the *coefficient of viscosity* of the medium or simply the *viscosity*.

The ratio $\nu = \eta/\varrho$, where, as always, ϱ is the density of the medium, is frequently encountered in problems of hydrodynamics and is called the *kinematic viscosity* of the medium.

We now find the dimensions of these quantities in the standard $\{L, M, T\}$ basis. Since $[\eta] = [FhS^{-1}v^{-1}]$, the dimension function corresponding to the viscosity in this basis has the form $\varphi_\eta = L^{-1}M^1T^{-1}$, and the dimension vector is $[\eta] = (-1, 1, -1)$. For the kinematic viscosity $[\nu] = [\eta/\varrho]$, therefore $\varphi_\nu = L^2M^0T^{-1}$ and $[\nu] = (2, 0, -1)$.

We now try to solve the previous problem on the drag force arising in the motion of the same ball, but now in a viscous medium. The initial dependence now looks like this: $F = f(\eta, \varrho, v, r)$. We analyse it in terms of dimensions:

	F	η	ϱ	v	r
L	1	−1	−3	1	1
M	1	1	1	0	0
T	−2	−1	0	−1	0

Clearly the vectors $[\varrho], [v], [r]$ are independent; $[F] = [\varrho] + 2[v] + 2[r]$ and $[\eta] = [\varrho] + [v] + [r]$. In view of the Π-Theorem we have

$$F = \varrho v^2 r^2 \cdot f(\mathrm{Re}^{-1},1,1,1), \tag{2.2}$$

where the function $f(\mathrm{Re}^{-1},1,1,1)$ remains unknown. This last function depends on the dimensionless parameter

$$\mathrm{Re} = \varrho v r / \eta = v r / \nu, \tag{2.3}$$

which plays a key role in questions of hydrodynamics.

This dimensionless quantity Re (the indication of the ratio of the force of inertia and the viscosity) is called the *Reynolds number* after the English physicist and engineer Osborn Reynolds, who first drew attention to it in his papers on turbulence in 1883. It turns out that as the Reynolds number increases, for example, as the speed of the flow increases or as the viscosity of the medium decreases, the character of the flow undergoes structural transformations (called bifurcations) evolving from a calm stable laminar flow to turbulence and chaos.

It is very instructive to pause at this juncture and wonder why the results (2.1) and (2.2) of the last two problems appear to be essentially the same. The wonder disappears if one considers more closely the variable quantity $f(\mathrm{Re}^{-1},1,1,1)$. Under the assumptions that, in modern terminology, are equivalent to the relative smallness of the Reynolds number, Stokes as long ago as 1851 found that $F = 6\pi\eta v r$. This does not contradict formula (2.2) but merely states that for small Reynolds numbers the function $f(\mathrm{Re}^{-1},1,1,1)$ behaves asymptotically like $6\pi\mathrm{Re}^{-1}$. In fact, substituting this value in formula (2.2) and recalling the definition (2.3) we obtain Stokes's formula.

2.7 Exercises

1. Since orchestras exist, it is natural to suggest that the speed of sound is weakly dependent (or not dependent) on the wave length ?

(Recall the nature of a sound wave, introduce the modulus of elasticity E of the medium and, starting from the dependence $v = f(\varrho, E, \lambda)$, prove that $v = c \cdot (E/\varrho)^{1/2}$.)

2. What is the law of the change of speed of propagation of a shock wave resulting from a very strong explosion in the atmosphere ?

(Introduce the energy E_0 of the explosion. The pressure in front of the shock wave can be ignored; the elasticity of the air no longer plays a role. Start by finding the law $r = f(\varrho, E_0, t)$ of propagation of the shock wave.)

3. Obtain the formula $v = c \cdot (\lambda g)^{1/2}$ for the speed of propagation of a wave in a deep reservoir under the action of the force of gravity. (Here c is a numerical coefficient, g is the acceleration of free fall and λ is the wave length.)

4. The speed of propagation of a wave in shallow water does not depend on the wave length. Accepting this observation as a fact show that it is proportional to the square root of the depth of the reservoir.

5. The formula used for determining the quantity of liquid flowing along a cylindrical tube (for example, along an artery) has the form

$$v = \frac{\pi \varrho P r^4}{8 \eta l},$$

where v is the speed of the flow, ϱ is the density of the liquid, P is the difference in pressure at the ends of the tube, r is the radius of cross-section of the tube, η is the viscosity of the liquid and l is the length of the tube. Derive this formula (to within a numerical factor) by verifying the agreement of the dimensions on both sides of the formula.

6. a) In a desert inhabited by animals it is required to overcome the large distances between the sources of water. How does the maximal time that the animal can run depend on the size L of the animal? (Assume that evaporation only occurs from the surface, the size of which is proportional to L^2.)

b) How does the speed of running (on the level and uphill) depend on the size of the animal? (Assume that the power developed and the corresponding intensity of heat loss (say, through evaporation) are proportional to each other, and the resistive force against horizontal motion (for example, air resistance) is proportional to the square of the speed and the area of the frontal surface.)

c) How does the distance that an animal can run depend on its size? (Compare with the answers to the previous two questions.)

d) How does the height of the jump of an animal depend on its size? (The critical load that can be borne by a column that is not too high is proportional to the area of cross-section of the column. Assume that the answer to the question depends only on the strength of the bones and the "capability" of the muscles (corresponding to the strengths of the bones).

Here we are dealing throughout with animals of size on the human scale, such as camels, horses, dogs, hares, kangaroos, jerboas, in their customary habitats. In this connection see the books by Arnold and Schmidt cited below.

7. After Lord Rayleigh, find the period of small oscillations of drops of liquid under the action of their surface tension, assuming that everything happens outside a gravitational field (in the cosmos).

(Answer: $c \cdot (\varrho r^3/s)^{1/2}$, where ϱ is the density of the liquid, r is the radius of the drop and s is the surface tension, $[s] = (0, 1, -2)$.)

8. Find the period of rotation of a double star. We have in mind that two bodies with masses m_1 and m_2 rotate in circular orbits about their common centre of mass. The system occurs in empty space and is maintained by the forces of mutual attraction between these bodies. (If you are puzzled, recall the gravitational constant and its dimension.)

9. "Discover" Wien's displacement law $\varepsilon(\nu,T)=\nu^3F(\nu/T)$ and also the Rayleigh-Jeans law $\varepsilon(\nu,T)=\nu^2T\;G(\nu/T)$ for the distribution of the intensity of black-body radiation as a function of the frequency and the absolute temperature.

[Wien's fundamental law (not the displacement law given above) has the form $\varepsilon(\nu,T)=\nu^3\exp(-a\nu/T)$ and is valid for $\nu/T\gg 1$, while the Rayleigh-Jeans law $\varepsilon(\nu,T)=8\pi\nu^2kT/c^3$ is valid for relatively small values of ν/T.

Both these laws (the specific intensity of radiation in the frequency interval from ν to $\nu+d\nu$) are united by Planck's formula (1900) launching the ground-breaking epoch of quantum theory:

$$\varepsilon(\nu,T)=\frac{8\pi}{c^3}\nu^2\frac{h\nu}{e^{h\nu/kT}-1}.$$

Here c is the velocity of light, h is Planck's constant, k is the Boltzmann constant ($k=R/N$, where R is the universal gas constant and N is Avogadro's number). Wien's law and the Rayleigh-Jeans law are obtained from Planck's formula for $h\nu\gg kT$ and $kT\gg h\nu$, respectively.]

Let ν_T be the frequency at which the function $\varepsilon(\nu,T)=\nu^3F(\nu/T)$ attains its maximum for a fixed value of the temperature T. Verify (after Wien) that we have the remarkable displacement law $\nu_T/T=\text{const}$. Find this constant using Planck's formula.

10. Taking the gravitational constant G, the speed of light c and Planck's constant h as the basic units, find the universal Planck units of length $L^*=(hG/c^3)^{1/2}$, time $T^*=(hG/c^5)^{1/2}$ and mass $M^*=(hc/G)^{1/2}$.

(The values $G=6.67\cdot 10^{-11}\text{H}\cdot\text{m}^2/\text{kg}^2$, $c=2.997925\cdot 10^8\text{m/s}$ and $h=6.625\cdot 10^{-34}\text{J}\cdot\text{s}$, other physical constants, as well as other information on units of measurement can be found in the books [4a], [4b], [4c].)

Many problems, analysed examples, instructive discussions and warnings relating to the analysis of dimensionality and principles of similarity can be found in the books [1], [2], [3], [4].

2.8 Concluding remarks

The little that has been said about the analysis of dimension and its applications already enables us to make the following observations.

The effectiveness of the use of the method mainly depends on a proper understanding of the nature of the phenomenon to which it is being applied. (By the way, in an early stage of analysis only people at the level of Newton, the brothers Bernoulli and Euler knew how to apply the analysis of infinitesimals without getting embroiled in paradoxes, which was required for extra intuition).[1]

Dimension analysis is particularly useful when the laws of the phenomenon have not yet been described. Namely, in this situation it sometimes reveals connections (albeit very general), which are useful for an understanding of the mechanism of the phenomenon and the choice of the direction of further investigations and refinements. We shall then demonstrate this by the example of Kolmogorov's approach to the description of the (still mysterious) fundamental phenomenon of turbulence.

The main postulate of dimension theory relates to the linear theory of similarity transformations, the theory of measurements, the notion of a rigid body and the homogeneity of a space, among other things. In Lobachevskii's hyperbolic geometry there are no similar figures at all, as is well known. Even so, locally this geometry admits a Euclidean approximation. Hence, as in all laws, the postulate of dimension theory is itself applicable in certain scales, depending on the problem. These scales were rarely known in advance and were most often discovered when incongruities arose.

The method shows that the larger the number of dimensionally independent quantities are, the simpler and more concrete the functional dependence of the quantities under study becomes. On the other hand, the more physical relations are discovered the less remains of the dimensionally independent quantities. (For example, distance can now be measured in light years.) So we see that the more we know, the less general dimension analysis gives us. Counterbalancing this, the penetration into essentially new areas is usually accompanied by the appearance of new dimensionally independent quantities (the algebraic aspect of dimensions and many other matters can be found, for example, in the book [14].)

Disregarding Problems 9 and 10 we restrict ourselves here to the discussion of phenomena described within the framework of the quantities of classical mechanics. This will suffice to begin with. But true enjoyment can only be obtained by reading the discussions of scholars, thinkers and, in general, professionals capable of a large-scale multischeme and unique view of the

---

[1] I quote the justified misgivings of V.I. Arnol'd concerning the possible overestimation of the Π-theorem: "Such an approach is extremely dangerous because it opens up the possibility of irresponsible speculation (under the name of dimension theory) in those places where the corresponding laws of similarity should be verified experimentally, since they do not at all follow from the dimensions of the quantities describing the phenomenon under study, and they are deep subtle facts". Rather the same relates to a clumsy use of multiplication tables, statistics or catastrophe theory.

Using these new publications of the present book I add that in his recent book "Mathematical understanding of nature" (MCCME, Moscow, 2009) discussing such a theory of adiabatic invariants V.I. (on p.117) observes that " The theory of adiabatic invariants is a strange example of a physical theory seemingly contradicting the purport of easily verifiable mathematical facts. In spite of such an undesirable property of this "theory" it provided remarkable physical discoveries to those who were not afraid to use its conclusions (even though they were mathematically unjustified)". In a word: "Think it out for yourself, solve it yourself, take it or leave it".

world or the subject matter. And this is in connection with various areas. It is like a symphony and it captivates!

[If you resign yourself to the "obscurantism" of dimension analysis but what has been set out still does not seem rather crazy, then it amuses me to quote the following excerpt from a well-known physicist (whom I shall not name so as not to accidentally subject his good name to attacks by less free-thinking people).

"Physicists begin the study of a phenomenon by introducing suitable units of measurement. It is unreasonable to measure the radius of an atom in metres or the speed of an electron in kilometres per hour; one needs to find appropriate units. There are already important immediate consequences of one such choice of units. Thus, from the charge e of an electron and its mass m one cannot form a quantity having the dimensions of length. This means that in classical mechanics the atom is impossible — an electron cannot move in a stationary orbit. The situation changed with the appearance of Planck's constant $\hbar$ ($\hbar = h/2\pi$). As is clear from the definition, $\hbar = 1{,}054 \cdot 10^{-34}$ J·s has dimensions of energy times time.[2] We can now form the quantity of the dimension of length: $a_0 = \hbar/me^2$.

If in this relation we substitute the values of the constants occurring therein, then we should get a quantity of the order of the dimensions of the atom; one obtains $0{,}5 \cdot 10^{-10}$m. Thus from a simple dimensional estimate one has found the size of the atom.

It is easy to see that $e^2/\hbar$ has the dimension of velocity, it is roughly 100 times smaller than the speed of light. If one divides this quantity by the speed of light c, then one obtains the dimensionless quantity $\alpha = e^2/\hbar c = 1/137$, characterizing the interaction of the electron with an electric field. This quantity is called the *fine structure constant*.

We have given estimates for the hydrogen atom. It is easy to obtain them for an atom with nuclear charge Ze. The motion of an electron in an atom is determined by its interaction with the nucleus, which is proportional to the product of the charge on the nucleus and the charge on the electron. Therefore for a nucleus with charge Ze, in the formulae for α and a_0 we must replace e^2 by Ze^2. In heavy elements with $Z \sim 100$ the velocity of the electrons is close to the speed of light."]

Finally we make some practical observations.

Dimension analysis is a good means of double checking:

a) if the dimensions of the left- and right-hand sides of an equation are not equal, then one must look for the error;

b) if under a sign that is not a degree function (for example, under a logarithm or exponential sign) there is a quantity that is not dimensionless, then one must look for the error (or one must look for a transformation getting rid of this situation);

[2] *Author's comment*: The dimension of h can be worked out from Planck's formula in Problem 9.

c) only quantities of the same dimension can be added.

(If $v = at$ is the velocity and $s = \frac{1}{2}at^2$ is the distance passed under uniform acceleration, then formally it is, of course, true that $v + s = at + \frac{1}{2}at^2$. However, from a physical point of view this equality reduces to two: $v = at$ and $s = \frac{1}{2}at^2$. Bridgeman, in whose cited book we gave this example, indicates a complete analogy with the equality of vectors , which gives rise to equalities of coordinates with the same name.)

Chapter 3
Further applications: hydrodynamics and turbulence

3.1 Equations of hydrodynamics (general information)

The basic classical characteristics of a moving continuous medium (liquid, gas) are, as is well known, velocity $v = v(x,t)$, pressure $p = p(x,t)$ and density $\rho = \rho(x,t)$ of the medium as a function of the point x of the region of flow and time t.

The analogue of Newton's equation $ma = F$ for the motion of an ideal continuous medium is Euler's equation

$$\rho \frac{dv}{dt} = -\nabla p. \tag{3.1}$$

If the medium is viscous, then on the right-hand side one must add to the force relating to the fall in pressure the force of internal friction; this gives us the equation

$$\rho \frac{dv}{dt} = -\nabla p + \eta \triangle v, \tag{3.2}$$

where η is the viscosity of the medium.

The above equation was introduced in 1827 by Navier in a certain special case and was then further generalized in turn by Poisson (1831), St. Venant (1843) and Stokes (1845). Since v is a vector field, this vector equation is equivalent to a system of equations for the coordinates of the field v. This system is called the Navier–Stokes system of equations, often called the NS-system.

For $\eta = 0$ we revert to Euler's equation relating to an ideal (non-viscous) fluid, for which one can ignore the loss of energy due to internal friction.

In addition to Euler's equation there is the so-called continuity equation

$$\frac{\partial \rho}{\partial t} + \operatorname{div}(\rho v) = 0, \tag{3.3}$$

expressing in differential form the law of conservation of mass (the change in the quantity of matter in any region of flow is the same as its flow through the boundary of this region).

For a homogeneous incompressible liquid $\rho \equiv \text{const}$, $\operatorname{div}(v) = 0$, the continuity equation holds automatically and the Navier–Stokes equation (3.2) takes the form

$$\frac{dv}{dt} = -\nabla \left(\frac{p}{\rho}\right) + \nu \triangle v, \tag{3.4}$$

V. Zorich, *Mathematical Analysis of Problems in the Natural Sciences*,
DOI 10.1007/978-3-642-14813-2_3, © Springer-Verlag Berlin Heidelberg 2011

where $\nu = \eta/\rho$ is the kinematic viscosity of the medium.

If the flow proceeds in the presence of mass forces (for example, in a gravitational field), then on the right-hand side of the NS-equation (3.2) the density f of these forces is added. Further, if we expand the total derivative on the left-hand side of the NS-equation bearing in mind that $\dot{x} = v$, then we obtain another way of writing the NS-equation:

$$\partial_t v + (v\nabla)v = \nu \triangle v + \frac{1}{\rho}(f - \nabla p). \tag{3.5}$$

If the flow is stationary, that is, the velocity field v is independent of time, then the last equation takes the form

$$(v\nabla)v = \nu \triangle v + \frac{1}{\rho}(f - \nabla p). \tag{3.6}$$

Here we do not intend to get further immersed in the enormous range of works relating to the Navier–Stokes equation. We merely recall that the following problem is included in the list of problems of the century (and a worthwhile prize has been set aside for its solution): if the initial and boundary conditions of the three-dimensional NS–equation (3.2) are smooth, then will the smoothness of the solution be preserved for ever or can a singularity spring up after a finite time ? (In the two-dimensional case no singularities arise.)

From the physical point of view there are probably other problems that present great interest, for example, the question how one can obtain from the NS–equation (3.2) a satisfactory description of turbulent flows and how the transition from turbulence to chaos occurs.

Thus, the equations of the dynamics of a continuous medium exist (for actual media they are supplemented by thermodynamic equations of state). In a number of cases these equations of the dynamics can be solved explicitly. In other cases it is possible to carry out the calculations of concrete flows on a computer. But, on the whole, a lot of further investigation is required, including, possibly, ideas new in principle.

Turning back to the very beginning, we now imagine that we do not yet know about the Navier–Stokes equations but are nevertheless interested in the flow of a continuous medium. For example, suppose that the flow of a homogeneous incompressible liquid having a velocity u at infinity runs into an object having a certain characteristic dimension l. We are interested in the stationary regime of the flow, that is, the resulting vector velocity field $v = v(r)$ as a function of the radius vector r of a point in space with respect to some fixed system of Cartesian coordinates. Let ϱ be the density of the liquid, η its viscosity and $\nu = \eta/\rho$ its kinematic viscosity.

Assuming that $v = f(r, \eta, \rho, l, u)$, we shall try to draw on a dimensional analysis:

	v	r	η	ρ	l	u
L	1	1	−1	−3	1	1
M	0	0	1	1	0	0
T	1	0	−1	0	0	−1

Hence, in view of the Π-theorem, we obtain the following relation between dimensionless quantities:

$$\frac{v}{u} = f(\frac{r}{l}, 1, \frac{\rho u l}{\eta}, 1, 1), \tag{3.7}$$

in which we find the Reynolds number $\mathrm{Re} := \frac{\rho u l}{\eta} = \frac{ul}{\nu}$.

We would have obtained the same result by supposing that $v = f(r, \nu, \rho, l, u)$, or if we had started with $v = f(r, \nu, l, u)$, in which the density is hidden in the kinematic viscosity ν.

Thus one can change the values of ρ, u, l, η; however, if at the same time we do not change their combination expressed by the Reynolds number Re, then the character of the flow remains the same to within the scale of the measurements of length and time (or length and velocity). Remarkable!

3.2 Loss of stability of the flow and comments on bifurcations in dynamical systems

The character of the flow for different values of the Reynolds number is, in general, different. As the Reynolds number increases there occur topological restructurings (bifurcations) of the flow. Its character changes from stable laminar flow to turbulence and chaos: for $\mathrm{Re} \asymp 10^0$ the flow is laminar; then for $Re \asymp 10^1$ the first critical value Re_1 appears and also the first bifurcation (first restructuring of the topology of the flow), and so on. The sequence $\mathrm{Re}_1 < \mathrm{Re}_2 < \ldots < \mathrm{Re}_n < \ldots$ of critical values rapidly converges, which, however, is a fairly universal phenomenon.

(This universality, which was discovered for the first time by M.J. Feigenbaum in 1978, bears his name *Feigenbaum universality*; it states that the following limit exists:

$$\lim_{n \to \infty} \frac{R_{n+1} - R_n}{R_n - R_{n-1}} = \delta^{-1},$$

where $\{R_n, n \in \mathbb{N}\}$ is the sequence of critical values at which the restructuring of the dynamical system (called bifurcation of period doubling) occurs, and $\delta = 4.6692\ldots$.) Thus the sequence of numbers Re_n has a limit Re_∞.

If the parameter Re of the problem is increased beyond the value Re_∞, then there occurs the regime, which in hydrodynamics is called turbulence. At very large values of the Reynolds number the motion becomes quite chaotic (as though it were an indeterminate random process).

At present the fundamental question remains open: can one arrive at a description of turbulence starting from the classical Navier–Stokes equations? Here we recall the words of Richard Feynman who, in another context, said (if I am not mistaken) that possibly Schrödinger's equation already contains the formula for life, but that does not rule out biology, which studies the living cell without waiting meanwhile to find out whether the existence of life is justified via Schrödinger's equation.

Having shown himself to be a true natural scientist A.N. Kolmogorov proposed (in 1941) a model for the development of turbulence, which, although it was subsequently refined, has remained basic and deserves special consideration. We have just chosen Kolmogorov's model to demonstrate the non-trivial application of dimensional analysis in the study of a phenomenon of which we have no fundamental description at our disposal.

Of course, the discovery of strange attractors (E.N. Lorenz, 1963) was a serious new general achievement of the theory of dynamical systems. This made it possible to concretize the ideology of the emergence of chaotic manifestations of a determinate system as its sensitivity to small changes in the initial conditions, and also to give a general-dynamical explanation of a phenomenon of turbulence (D. Ruelle & F. Takens, 1971) (See also the commentary in the paper [5]).

3.3 Turbulence (initial ideas)

Introducing the collection [9] of articles on turbulence, Academician O.M. Belotserkovskiĭ recalled that when he studied at the physics faculty in Moscow University, lectures on electricity were given to him by Professor S.G. Kalashnikov, who at the very first lecture related the following. Once in an examination he (Kalashnikov) asked a student "What is electricity?" The student began to fidget and fuss and replied "Oh dear, I knew it yesterday but now I have forgotten". To this Kalashnikov observed "There was only one man who knew it, and even he forgot it!"

The situation with turbulence is roughly the same, although people of various disciplines have speculated about this, of course primarily physicists, mathematicians and astronomers.

A fast river flowing past the pier of a bridge forms a vortex, which as it percolates forms patterns drawn by Leonardo da Vinci, who penetrated everything with his thought and eye. Vortices can also be observed in the air in clouds of dust behind a whirling machine or, much more pleasingly simply in clouds capriciously deforming themselves in front of one's eyes. Cosmic vortices form galaxies. Water from a tap stops flowing peacefully when the tap is switched on too violently. A small aircraft is not allowed to take off behind a large liner. And through the porthole of an aeroplane one observes with interest the seemingly minute (as seen from above) ocean ships behind which there trails a clearly distinguishable turbulent wake.

In science the term *turbulence* was established by the end of the nineteenth century, after Maxwell, Lord Kelvin and Reynolds, although already Da Vinci made good use of it.

3.4 The Kolmogorov model

3.4.1 *The multiscale property of turbulent motions*

As above, we consider the flow round a body, that is, a flow running into a body of characteristic dimension l. If the velocity of the flow is large or the viscosity of the medium is small, then for very large values of the Reynolds number a region of turbulence is developed in some volume behind the body. In that region the flow is extremely unstable, chaotic and has the character of pulsations of different scales propagated in the region of turbulence. The change in the velocity field $v = v(x,t)$ is reminiscent here of a stationary random process when not the velocity field is stationary, but only certain averaged probabilistic characteristics (expressed, for example, by histograms of the probability distribution of some or other quantities connected with the flow).

Kolmogorov observed that for very large values of the Reynolds number, in the region of turbulence the picture of the flow, was locally homogeneous and isotropic, although still complicated.

Turbulence can (or should) be regarded as a manifestation of the interaction of the motions of different scales. On the motion of large scales pulsations of smaller scales are imposed and they are transferred by the motions of the larger scales (train passengers moving in the restaurant car participate in the motions of scales of the distance between towns, but can be looked at in the scales of the restaurant car as well).

Let us explain the idea of the multiscale property. A cell lives its own life. Associations of cells interacting with each other form a certain tissue. A group of tissues form an organ. A group of organs form an organism. An organism sits in a machine and goes to work. On the streets of a town a transport flow forms. All these flows together with the towns and countrysides are carried in space by a rotating Earth, and so on. Even closer to our theme could be the example relating to the multiscale life of the ocean or the atmosphere.

When we talk about the movement of a passenger, then it is clear that we have in view the characteristics of his motion within the scales of the restaurant car. We do not single out an individual passenger when we talk about speed in the scales of the motion of the entire train.

Turbulence is primarily the relative motions in the locality of the fluid and not the absolute transfer of motion in which it participates as an element of a larger-scale motion.

If this is the case, then in turbulence we have in front of us an entire spectrum of motions of different scales and we want, for example, to indicate some characteristic parameters of the motions of different scales: distribution of the energy of turbulent flow with respect to the scales of motion, the relative velocities of motions of different scales, the velocity of dispersion of particles in a turbulent flow, and so on.

3.4.2 *Developed turbulence in the inertia interval*

We now turn to a more concrete discussion.

Consider a flow, for example, (as above) round a body of characterisic dimension l for large values of the Reynolds number (Re $\gg 1$). For Re $\gg 1$ the turbulence that arises is usually called *developed turbulence*. We shall suppose, after Kolmogorov, that the developed turbulence in motions of scales $\lambda \ll l$ and far from solid walls (that is, at a distance much larger than the size of λ) is isotropic and homogeneous.

The condition Re $\gg 1$ can be treated as smallness of viscosity. Viscosity only manifests itself in small-scale motions of some scale λ_0 since internal friction only occurs between close particles of the liquid. When one goes over to large masses of the liquid, the viscosity is insignificant since in this case the dissipation of energy is negligibly small by comparison with the kinetic energy of the inertial motion of the large mass.

The interval of scales $\lambda_0 \ll \lambda \ll l$ is called the *inertia interval*. In motions of these scales the viscosity can be ignored. The quantity λ_0 is called the *inner scale of the turbulent motion* and l is called the *outer scale of the turbulent motion*.

3.4.3 *Specific energy*

The steady turbulent regime is maintained by the expenditure of external energy dispersing in the liquid due to its viscosity. Let ε be the specific power of dissipation, more precisely, the amount of energy dissipated by a unit mass of the liquid in unit time. In accordance with this definition the quantity ε has the following dimension vector in the standard basis $\{L, M, T\}$: $[\varepsilon] = (2,0,-3)$. [The force $F = ma$ has dimension (1, 1, -2). Energy, work and potential energy $F \cdot h$ have dimension (2, 1, -2). Hence $[\varepsilon] = (2,0,-3)$.]

The kinetic energy of a flow running at a velocity u is decreased as a result of the dissipation of energy due to the internal friction in the viscous liquid.

The change Δu in the average velocity of the basic motion occurs in a spread of order l (up to the encounter with the body and beyond)

The quantity ε must be determined by this loss of kinetic energy of the basic motion, that is, it must be a function $\varepsilon = f(\rho, \Delta u, l)$. We then conclude on the basis of the Π-theorem that this quantity is of the order

$$\varepsilon \sim \frac{(\Delta u)^3}{l}. \tag{3.8}$$

Similarly we obtain the following equation for the fall in pressure:

$$\Delta p \sim (\Delta u)^2 \rho. \tag{3.9}$$

3.4.4 *Reynolds number of motions of a given scale*

Since we are interested in motions of different scales, we can associate with each scale λ the Reynolds number corresponding to it

$$Re_\lambda := \frac{v_\lambda \cdot \lambda}{\nu}. \tag{3.10}$$

In terms of these the inner scale of turbulence, that is, the quantity λ_0, must be determined by the condition that the Reynolds number have the order $\mathrm{Re}_{\lambda_0} \sim 1$, since a greater value of the Reynolds number would be equivalent to a small viscosity.

3.4.5 *The Kolmogorov–Obukhov law*

We now find the average velocities v_λ of motions of the scale λ (or, which is the same, the change of the average velocity of a turbulent flow in a spread of distances of order λ). In the inertia interval, when $\lambda_0 \ll \lambda \ll l$, we can assume that $v_\lambda = f(\rho, \varepsilon, \lambda)$. Then, on the basis of the Π-theorem we conclude that

$$v_\lambda \sim (\varepsilon \lambda)^{1/3}. \tag{3.11}$$

This relation is called the *Kolmogorov–Obukhov law*. (A.M. Obukhov, who was student of Kolmogorov during the 1940s subsequently became an academician and director of the Institute of Physics of the Atmosphere in Moscow. Another student of Kolmogorov at the same time was A.S. Monin, who also became an academician and was director of the Institute of Oceanography. About this Kolmogorov joking said that one of his students is in charge of the ocean and another is in charge of the atmosphere.)

3.4.6 *Inner scale of turbulence*

We now find the inner scale λ_0 of turbulent flow. As we know, we need to find it from the condition that $\mathrm{Re}_{\lambda_0} \sim 1$.

For the Reynolds number Re for the basic motion of scale l, in accordance with the general definition of Reynolds number and in accordance with formula (3.10), we have $\mathrm{Re} \sim (\Delta u \cdot l)/\nu$. Taking into account relations (3.8) and (3.11) we get

$$\mathrm{Re}_\lambda \sim \frac{v_\lambda \cdot \lambda}{\nu} \sim \frac{(\varepsilon\lambda)^{1/3}\lambda}{\nu} \sim \frac{\Delta u \cdot (\lambda)^{4/3}}{\nu l^{1/3}} = \mathrm{Re}\left(\frac{\lambda}{l}\right)^{4/3}.$$

Setting $\mathrm{Re}_\lambda \sim 1$ we find that

$$\lambda_0 \sim \frac{l}{\mathrm{Re}^{3/4}}. \tag{3.12}$$

For the corresponding velocity we have

$$v_{\lambda_0} \sim (\varepsilon\lambda_0)^{1/3} \sim \frac{\Delta u}{l^{1/3}} \cdot \frac{l^{1/3}}{\mathrm{Re}^{1/4}} = \frac{\Delta u}{\mathrm{Re}^{1/4}}. \tag{3.13}$$

3.4.7 *Energy spectrum of turbulent pulsations*

We associate with the scale of length λ as a wave length the number $k := 1/\lambda$. Let $E(k)dk$ be the kinetic energy in the pulsations (motions) with wave number k in the interval dk referring to unit mass of the liquid.

We shall find the density $E(k)$ of this distribution. Since $E(k)dk$ has the dimension of energy relative to unit mass and $[dk] = (-1,0,0)$, we find that $[E(k)] = (3,0,-2)$. Combining ε and k and using Kolmogorov's dimensional arguments we obtain

$$E(k) \sim \varepsilon^{2/3} k^{-5/3}. \tag{3.14}$$

Assuming that v_λ determines the order of magnitude of the kinetic energy of the motions of all scales not exceeding λ we can again obtain the Kolmogorov–Obukhov law

$$v_\lambda^2 \sim \int_{k=1/\lambda}^{\infty} E(k)dk \sim \varepsilon^{2/3} k^{-2/3} \sim (\varepsilon\lambda)^{2/3}$$

and $v_\lambda \sim (\varepsilon\lambda)^{1/3}$.

3.4.8 *Turbulent mixing and dispersion of particles*

Two particles situated in a turbulent flow at a mutual distance λ apart become separated to a distance $\lambda(t)$ over an interval of time t. We shall find the speed $\lambda'(t)$ of separation of the particles. As in the derivation of the Kolmogorov–Obukhov law, we assume that $\lambda' = f(\rho, \varepsilon, \lambda)$, and in complete accordance with formula (3.11) we obtain

$$\frac{d\lambda}{dt} \sim (\varepsilon\lambda)^{1/3}. \tag{3.15}$$

As is clear, the speed of separation increases as λ increases. This is explained by the fact that in the process under consideration only motions of scale less than λ participate. The larger-scale motions transfer the particles but do not lead to their separation.

Part II
Multidimensional Geometry and Functions of a Very Large Number of Variables

Introduction

Almost all the bulk of a multidimensional body is concentrated at its boundary. For example, if one removes from a 1000-dimensional watermelon of radius one metre the peel of thickness 1 centimetre, then there remains less than one thousandth of the entire watermelon.

This phenomenon of localization or concentration of the measure has numerous unexpected manifestations. For example, any more-or-less regular function on a multidimensional sphere is almost constant in the sense that if one takes randomly and independently a pair of points of the sphere and calculates the values of the function at these points, then with high probability they will turn out to be almost the same.

From the point of view of a mathematician used to dealing with functions of one, two or several (but not a great number of) variables this may appear implausible. But in fact this ensures the stability of the basic parameters within our habitat (temperature, pressure, and so on), it lies in the foundations of statistical physics, is studied in probability theory under the name of the law of large numbers and has many applications (for example, in the transmission of information along a communication channel in the presence of noise).

The phenomenon of the concentration of measure explains in some respect both the statistical stability of the values of thermodynamic quantities which gave rise to the Boltzmann ergodic hypothesis and the remarkable ergodic theorems that arose with the aim of justifying it.

The principle of concentration is set forth in Chapter 2, which can be read independently of Chapter 1, where we give examples of areas in which functions of a large number of variables appear in a natural way.

In Chapter 1 (which, of course, can also be read independently) we dwell in detail on a less popular example — the transmission of information along a communication channel. We introduce and discuss the sampling theorem — Kotel'nikov's formula – the basis of modern digital representation of a signal. In Chapter 3 we supplement these discussions with Shannon's theorem on the speed of transmission along a communication channel in the presence of noise.

Chapter 1
Some examples of functions of very many variables in natural science and technology

1.1 Digital sampling of a signal (CIM —code-impulse modulation)

1.1.1 *The linear device and its mathematical description (convolution)*

A good mathematical model of many devices and instruments is the linear operator. The term "linear device with time-invariant properties" in mathematical language means that it is a linear operator A acting on functions of time and commuting with the shift operator T, that is, $AT = TA$, where $T = T_\tau$, and $(T_\tau f)(t) = f(t - \tau)$.

For example, if A is a record player, then tomorrow it should produce from the same disc the same music as today, only with the natural shift in time. In accordance with the accepted radio-engineering terminology, the function f on which the operator acts is called the *signal*, more precisely, the *input signal* or *input*, while the result Af is called the *signal at the output* or *output* and is denote by $\tilde{f}$.

Since a continuous function f can be well approximated by a step function, it is easy to see that if one knows the response of such a device A to an elementary step-like datum then one can find its response to any input signal f.

Ideally the step-like datum is converted to the unit impulse, the δ-function. If we write down the identity $f(t) = \int f(\tau)\delta(t-\tau)d\tau$, then we immediately see that $Af(t) = \int f(\tau)A\delta(t-\tau)d\tau = \int f(\tau)\tilde{\delta}(t-\tau)d\tau =: f * \tilde{\delta}$, where $*$ is the symbol for the operation of convolution of functions. Hence $Af = f * \tilde{\delta}$.

The function $A\delta = \tilde{\delta}$, that is, the response of the device to the unit impulse (δ-function) is called the *instrumental function of the device* and is often denoted by the symbol E. Thus, from a mathematical point of view the device A is simply the convolution operator $Af = f * \tilde{\delta} = f * E$. Thus the solution of convolution equations has very specific direct applications (for example, the recovery of a transmitted signal from the received signal $Af = \tilde{f}$).

V. Zorich, *Mathematical Analysis of Problems in the Natural Sciences*,
DOI 10.1007/978-3-642-14813-2_4, © Springer-Verlag Berlin Heidelberg 2011

1.1.2 Fourier-reciprocal (spectral) description of a linear device

We recall that the frequency ν of a periodic process is usually measured by the number of complete cycles in unit time (one Hertz is one complete oscillation in one second; it is denoted by 1Hz). The angular or circular frequency $\omega = 2\pi\nu$ differs from the frequency ν merely by the factor 2π corresponding to measurement in radians per unit time.

We shall calculate the response $\tilde{f}$ of the device to an input signal $f = e^{i\omega t}$ (hence, by Euler's formula $e^{i\omega t} = \cos\omega t + i\sin\omega t$ we also know the response of the device to simple harmonic oscillation $\sin\omega t$ of angular frequency ω; here it is convenient to use complex language):

$$Af = f * \tilde{\delta} = f * E = \int f(t-\tau)\tilde{\delta}(\tau)d\tau =$$
$$\int e^{i\omega(t-\tau)}E(\tau)d\tau = \left(\int E(\tau)e^{-i\omega\tau)}d\tau\right)e^{i\omega t} = P(\omega)e^{i\omega t}.$$

We have obtained an oscillation with the same frequency as the input, but, possibly, a change in amplitude by a factor $|P(\omega)|$ and a change of phase corresponding to $\arg P(\omega)$. The quantity P as a function of ω is called the *spectral characteristic of the device*. It is clear that the spectral characteristic of the device is (to within a normalizing factor) the Fourier transform $\hat{E}$ of the instrumental function E of this device: $P = 2\pi\hat{E}$. We stipulate that $p(\nu) := P(2\pi\nu) = P(\omega)$.

Recall that in terms of the frequencies ω and ν the Fourier transforms $\hat{f}$ and $\check{f}$, respectively, and the Fourier integral of the function f (formula for the inverse Fourier transform in L_2) have the form

$$f(t) = \int \hat{f}(\omega)e^{i\omega t}d\omega, \qquad \text{where} \quad \hat{f}(\omega) = \frac{1}{2\pi}\int f(t)e^{-i\omega t}dt;$$

$$f(t) = \int \check{f}(\nu)e^{i2\pi\nu t}d\nu, \quad \text{where} \quad \check{f}(\nu) = \int f(t)e^{-i2\pi\nu t}dt.$$

Since the Fourier transform is invertible, the function E can be recovered from the function P (or p). Hence the spectral characteristic or the spectral function P (or p) of the device as well as its instrumental function E completely determines the device A.

We then calculate Af knowing P. Representing f by a Fourier integral we find the representation Af in the form of a Fourier integral:

$$f(t) = \int \hat{f}(\omega)e^{i\omega t}d\omega \quad \text{and} \quad Af(t) = \int \hat{f}P(\omega)e^{i\omega t}d\omega.$$

In particular, if $f = \delta$, then

$$E(t) = \tilde{\delta}(t) = A\delta(t) = \frac{1}{2\pi} \int P(\omega) e^{i\omega t} d\omega \quad \left(= \int \hat{E}(\omega) e^{i\omega t} d\omega \right).$$

1.1.3 *Functions and devices with a compactly supported spectrum*

The devices that we actually have to deal with, like hearing and vision, are capable of receiving a signal only in a certain range of frequencies. Therefore it is natural to deal with functions and devices with a compactly supported spectrum.

If the spectrum (Fourier transform $\hat{f}$) of f is compactly supported (identically zero outside some compact set), then, in its representation as a Fourier integral, the integral is between finite limits:

$$f(x) = \int_{-a}^{a} \hat{f}(\omega) e^{i\omega x} d\omega.$$

In view of their importance, functions with a compactly supported spectrum have been the subject of independent mathematical investigation.

If a function g belongs to the space $L_2(\mathbb{R})$, then, as is known from the theory of the Fourier transform, the function

$$f(x) = \int_{-a}^{a} g(\omega) e^{i\omega x} d\omega$$

also belongs to the space $L_2(\mathbb{R})$. Moreover, by applying the Cauchy–Schwarz–Bunyakovskiĭ inequality, it is easy to see that it is bounded on the real axis and can be extended to the whole of the complex plane as an entire function; furthermore, $|f(x+iy)| \leq ce^{a|y|}$. The class of entire functions of this form is called the Wiener class and is denoted by W_α (see, for example, [4]).

The Paley–Wiener theorem states that $f \in W_a$ if and only if this function admits a representation

$$f(z) = \int_{-a}^{a} g(\omega) e^{i\omega z} d\omega,$$

where $g \in L_2(\mathbb{R})$.

1.1.4 *The ideal filter and its instrumental function*

We now turn to a device with a compactly supported spectrum. Here the simplest basic example is a device whose spectral function $P(\omega)$ is equal to unity on the interval $[-a,a] = [-\Omega,\Omega]$ and is equal to zero outside this interval

Such a device, as we now understand, will transmit without distortion all harmonics with frequency not greater than Ω ($|\omega| \leq \Omega = a$) and will not react to higher frequencies. We call such a device a *low-frequency filter with upper boundary frequency* Ω.

We find the instrumental function of this low-frequency filter with upper boundary frequency a (normalized with an averaging factor):

$$E_a(t) = \frac{1}{2a}\int_{-a}^{a} e^{i\omega t}d\omega = \frac{\sin at}{at}.$$

The importance of this function in electrotechnology and the theory of transmission along a communication channel has led to the appearance of the following special notation:

$$\operatorname{sinc} x := \frac{\sin x}{x}.$$

Some explanation why the letter c has suddenly appeared at the end of the above word will be given below.

1.1.5 *The sampling theorem (Kotel'nikov–Shannon formula)*

First we carry out the following calculation in its simplest version so as not to obscure the essentials with unwanted details. We take a regular function f with compactly supported spectrum $\hat{f} = \frac{1}{2\pi}\phi$ concentrated on the interval $[-\pi,\pi]$, expand ϕ in a Fourier series and calculate the coefficients:

$$f(t) = \frac{1}{2\pi}\int_{-\pi}^{\pi}\phi(x)e^{itx}dx;\ \phi(x) = \sum_{-\infty}^{\infty} c_n e^{-inx};\ c_n = \frac{1}{2\pi}\int_{-\pi}^{\pi}\phi(x)e^{inx}dx = f(n).$$

In the first integral we substitute for the function ϕ its Fourier series and integrate term by term. Assuming that ϕ is "good" we get

$$f(t) = \sum_{-\infty}^{\infty} c_n\left(\frac{1}{2\pi}\int_{-\pi}^{\pi} e^{ix(t-n)}dx\right) = \sum_{-\infty}^{\infty} f(n)\frac{\sin\pi(t-n)}{\pi(t-n)} = \frac{\sin\pi t}{\pi}\sum_{-\infty}^{\infty} f(n)\frac{(-1)^n}{t-n}.$$

By complete analogy for a general function with compactly supported spectrum

$$f(t) = \frac{1}{2a}\int_{-a}^{a}\phi(x)e^{itx}dx,$$

we now expand ϕ in a Fourier series on the interval $[-a,a]$; we find that

$$\phi(x) = \sum_{-\infty}^{\infty} c_n e^{-i\frac{\pi}{a}nx}, \quad \text{where} \quad c_n = \frac{1}{2a}\int_{-a}^{a} \phi(x) e^{i\frac{\pi}{a}nx} dx = f\left(\frac{\pi}{a}n\right),$$

and obtain the following representation of a function with compactly supported spectrum, known as the *Kotel'nikov formula* or the *Kotel'nikov–Shannon formula* or the *sampling theorem*:

$$f(t) = \sum_{-\infty}^{\infty} f\left(\frac{\pi}{a}n\right) \frac{\sin a(t - \frac{\pi}{a}n)}{a(t - \frac{\pi}{a}n)}.$$

This formula was obtained by Kotel'nikov [1] in 1933 and rediscovered by Shannon [2] in 1949. As an interpolation formula (special case of Lagrange's formula for entire functions of finite degree) it was already known to mathematicians (see, for example, [4] and the literature cited therein). The merit of the Kotel'nikov–Shannon contribution consists in the interpretation of this formula from the point of view of signal coding and transmission of information along a communication channel.

The formula can, of course, be converted to the form

$$f(t) = \frac{\sin at}{a} \sum_{-\infty}^{\infty} f\left(\frac{\pi}{a}n\right) \frac{(-1)^n}{t - \frac{\pi}{a}n},$$

however the original formula is more important for us at the moment.

1.1.6 Code-impulse modulation of a signal (CIM)

The Kotel'nikov–Shannon formula shows that a regular signal with a compactly supported spectrum of frequencies lying in the interval $[-a, a]$ can be completely recovered from the set of its discrete values that can be read off in the interval $\Delta = \frac{\pi}{a}$ (hence the name "sampling theorem").

Furthermore, the signal is recovered via the combination consisting of signals of the instrumental function of the simplest low-frequency filter with the upper boundary frequency a, that is, this is still a relatively simply technically realizable formula for discrete coding and transmission of the signal. (In principle, for example, a periodic signal can be given by its Fourier series, while an analytic function can be given by the coefficients of its Taylor expansion, although this is not always convenient from the point of view of the possibility of an instrumental realization.)

The idea described above of discrete coding of a signal lies at the foundation of modern digital technology of write signals, storage and reproduction of information (music, video, libraries, retrieval systems, and so on).

We now observe that the special notation $\operatorname{sinc} x := \frac{\sin x}{x}$ must be interpreted as "sine-counting": the letter "c" standing for "counting".

1.1.7 *Transmission capacity of an ideal communication channel*

The theoretical development of the notion of information and its quantitative description was stimulated by the emergence at the beginning of the 20th century of the telegraph and the radio signal. Even before the appearance of Kotel'nikov's formula specialists were groping about for approaches to the solution of these problems; see, for example, the collection of articles [3]. Thus, in honour of Nyquist the sampling interval $\Delta = \frac{\pi}{a}$ is (not without reason) called the *Nyquist interval* (apparently after Shannon). In time Δ one sample value of the function f is transmitted, that is, Δ^{-1} is the number of sample values transmitted in unit time (second).

Although we do not know at the time either the notion of information or how one can measure it, the Kotel'nikov–Shannon formula gives us the connection between the width of the usable band of frequencies and the speed of transmission of information along the communication channel. But we shall talk more about this below.

1.1.8 *Evaluation of the dimension of a TV signal*

Suppose, for example, that (as Shannon did) we consider a TV-signal having a frequency of $W = 5$ MHz (1 megahertz $= 10^6$ Hertz) and duration 1 hour. We shall calculate the length of the vector of sample values corresponding to this signal, that is, we find the number $N = T/\Delta = 2WT$ of sample values: $N = 2 \cdot 5\text{MHz} \cdot 1\text{h} = 2 \cdot 5 \cdot 10^6 \cdot 60^2 = 3{,}6 \cdot 10^{10}$. This is a vector in a space $\mathbb{R}^N$ of enormous dimension. The geometry of such a space has its own peculiarities. This is mainly the topic of Chapter 2.

1.2 Some other areas of multiparameter phenomena and spaces of large dimension

Of course, in order to understand that the digital representation of some complex signal or information requires a large number of symbols, one does not need to know everything set forth above, including the Koltel'nikov–Shannon formula. We dwelt on this in detail because here quite interesting things can be discussed which require of the reader hardly any preliminary preparation. On the other hand, this information is not all that often known in mathematical circles. This is why we intend merely to mention below other (even more fundamental) areas where the quantities and functions being described also usually depend on a very large number of parameters, that is, essentially they too relate to a space of large dimension.

1.2.1 *Molecular theory of matter*

The thermodynamic functions, for example, the pressure of a gas, from the point of view of statistical physics rather than phenomenological thermodynamics, depends on an enormous number of variables. (Recall, for instance, Avogadro's number $N_A = 6.022 \cdot 10^{23}\text{mole}^{-1}$.)

Surprisingly (at least before the explanations given below relating to multidimensional geometry and the principle of concentration) in this connection (then explained that thanks to this) the values of these functions remain stable. Below we shall be concerned with kinetic theory and shall even in fact obtain the famous Maxwell's distribution.

1.2.2 *Phase space in classical Hamiltonian mechanics*

The phase space of some complicated mechanical system is, as a rule, multidimensional. Hence many functions related to the system turn out to be functions of a large number of variables

1.2.3 *The Gibbs thermodynamic ensembles*

Combining the ideas of thermodynamics and Hamiltonian mechanics Gibbs introduced into statistical physics a remarkable mathematical structure — the Hamiltonian system endowed with a measure evolving (sometimes to equilibrium which is of particular interest) under the action of a Hamiltonian flow in the phase space of the system.

Statistical physics (Boltzmann's work) and Newton's three-body and many-body problem (Poincaré's work) started the appearance and rapid development of the mathematical theory of dynamical systems and its numerous diverse problems, some of which are still unsolved. (For example, the questions of the description of the phase transitions, turbulence and chaos.)

1.2.4 Probability theory

Of course, large numbers (of trials), their averaging interactions, the characterization of sparse noticeable deviations, and so on form the foundation of the philosophy of ideas and the core of a whole area of mathematics, namely, probability theory.

Below we shall for the moment be concerned with probability theory by considering (as an example of the application of the geometric considerations) the derivation of Gauss's normal distribution for handling errors of observations.

Chapter 2
Concentration principle and its applications

2.1 The ball and sphere in Euclidean space $\mathbb{R}^n$ with $n \gg 1$

2.1.1 Concentration of the volume of a ball as $n \to \infty$.

Consider a ball $B^n(r)$ of radius r in Euclidean space $\mathbb{R}^n$ of large dimension n. Let $\operatorname{Vol} B^n(r)$ be its volume. Then

$$\frac{\operatorname{Vol} B^n(r+\Delta)}{\operatorname{Vol} B^n(r)} = \frac{(r+\Delta)^n}{r^n} = \left(1+\frac{\Delta}{r}\right)^n .$$

This means that if the dimension n is large, then one only has to increase the radius of the ball by $\Delta = \frac{1}{n}r$ so as to more than double its volume.

For example, if in $\mathbb{R}^{1000}$ a watermelon of radius 1m has a peel of thickness 1cm, then after the peel is removed there remains less than one thousandth of its volume. Thus the overwhelming part of the volume of a multidimensional ball is concentrated in a small neighbourhood of its boundary.

2.1.2 Thermodynamic limit

As is well known, Maxwell in his preoccupation with statistical physics, more precisely with kinetic theory, was the first to discover the law (*Maxwell's law*) of the distribution of molecules of a given volume of gas with regard to their speed and, as a consequence, with regard to their kinetic energy.

Suppose that in a given volume at a given temperature there are n molecules of mass m of the gas; let v_i be the speed of the ith molecule, and E_n the total kinetic energy of the molecules. As n increases E_n increases and has order n: $E_n \asymp n$. In formal notation this means that

$$\frac{1}{2}mv_1^2 + \dots + \frac{1}{2}mv_n^2 = E_n; \qquad \sum_{i=1}^{n} v_i^2 = \frac{2E_n}{m} \asymp n.$$

V. Zorich, *Mathematical Analysis of Problems in the Natural Sciences*,
DOI 10.1007/978-3-642-14813-2_5,

We shall seek a statistical characterization of this ensemble of particles as $n \to \infty$ subject to the condition that $E_n \asymp n$. The passage to the limit under this condition is called the *thermodynamic passage to the limit* or more concisely but less precisely, the *thermodynamic limit*. (The precise definitions and a modern mathematical treatment of thermodynamic limits can be found in the book [9].)

From the mathematical point of view, here we are dealing with a $(3n-1)$-dimensional sphere in $\mathbb{R}^{3n}$ whose radius increases as $n^{1/2}$ as $n \to \infty$.

In the statistical study of the law of the distribution of independent errors Δ_i of measurement, in the derivation of the Gauss law we suppose that the variance D is finite and we have the absolutely similar situation:

$$\frac{1}{n}(\Delta_1^2 + \ldots + \Delta_n^2) \asymp D; \qquad \sum_{i=1}^{n} \Delta_i^2 \asymp n.$$

If the molecules as well as the observational errors are assumed to be on an equal footing and the points of the sphere corresponding to them are uniformly distributed along its surface, then both the statistics of the molecules of the gas and the statistics of the errors of measurement result in the statistics of the projection of an $(n-1)$-sphere onto a line in $\mathbb{R}^n$ as $n \to \infty$, when the radius of the sphere increases as $n^{1/2}$.

We make what we have just said more precise and carry out the corresponding calculations below.

2.1.3 *Concentration of the area of a sphere*

In Euclidean space $\mathbb{R}^n$ consider a hypersphere $S^{n-1}(r)$ of radius r and with centre at the origin. We introduce in $\mathbb{R}^n$ spherical coordinates, counting the angle ψ from the positive direction of (what we shall call) the x-axis. Hence, $x = r\cos\psi$, $dx = -r\sin\psi d\psi$ and $\sin\psi = (r^2 - x^2)^{1/2}/r = (1 - (\frac{x}{r})^2)^{1/2}$.

The area of an elementary spherical shell corresponding to the angular interval $(\psi, \psi + d\psi)$ is given by the formula

$$\begin{aligned}\sigma_{n-2}(r\sin\psi) r d\psi = c_{n-2}(r\sin\psi)^{n-2} r d\psi = c_{n-2} r^{n-2} \sin^{n-3}\psi \; r\sin\psi d\psi = \\ = c_{n-2} r^{n-2}(1 - (x/r)^2)^{\frac{n-3}{2}}(-dx). \quad (2.1)\end{aligned}$$

Here $\sigma_{n-2}(\rho)$ is the area of a $(n-2)$-sphere of radius ρ and $c_{n-2} = \sigma_{n-2}(1)$.

From formula (2.1) we now find the area of the spherical shell projected onto the interval $[a,b] \subset [-r,r]$ of the x axis:

$$c_{n-2} r^{n-2} \int_a^b \left(1 - (x/r)^2\right)^{\frac{n-3}{2}} dx. \qquad (2.2)$$

The ratio of this area to the entire area of the sphere $S^{n-1}(r)$ of radius r is equal to

$$P_n[a,b] := \frac{\int_a^b (1-(x/r)^2)^{\frac{n-3}{2}}\,dx}{\int_{-r}^r (1-(x/r)^2)^{\frac{n-3}{2}}\,dx}. \tag{2.3}$$

Passing to the thermodynamic limit as $n \to \infty$ and $r = \sigma\, n^{1/2}$ we obtain the normal distribution

$$P_n[a,b] := \frac{\int_a^b e^{-\frac{x^2}{2\sigma^2}}\,dx}{\int_{-\infty}^{\infty} e^{-\frac{x^2}{2\sigma^2}}\,dx}. \tag{2.4}$$

As already pointed out above, it simultaneously shows up as the Gauss distribution in the statistical theory of handling the results of observations (theory of errors) and as the Maxwell distribution in statistical physics (kinetic theory). Also involved is the central limit theorem of probability theory.

If we fix r and let n tend to infinity, then the quantity $P_n[a,b]$ in formula (2.3) will tend exponentially rapidly to zero for $0 < a < b \le r$. We now substantiate this calculation.

First we recall the classical results relating to the asymptotic behaviour of the Laplace integral

$$F(\lambda) := \int_{I=[a,b]} f(x)e^{\lambda S(x)}\,dx.$$

Suppose that both functions f and S are defined and regular on the interval I of integration and that S is real and has a single absolute maximum on I which is attained at the point $x_0 \in I$ and $f(x_0) \neq 0$.

Then as $\lambda \to +\infty$ the asymptotic behaviour of the Laplace integral is the same as if the integral extended only in an arbitrarily small neighbourhood of the point x_0 in I (this is the so-called *localization principle*).

After this localization the question reduces to a consideration of the special case when $I = [x_0, x_0+\varepsilon]$ and/or $I = [x_0-\varepsilon, x_0]$ with ε an arbitrarily small positive number. Using the Taylor expansion we now find that as $\lambda \to +\infty$

$$F(\lambda) = \frac{f(x_0)}{-S'(x_0)} e^{\lambda S(x_0)} \lambda^{-1} \left(1 + O(\lambda^{-1})\right) \tag{A}$$

if $x_0 = a$ and $S'(x_0) \neq 0$ (hence $S'(x_0) < 0$, since $a < b$);

$$F(\lambda) = \sqrt{\frac{\pi}{-2S''(x_0)}} f(x_0) e^{\lambda S(x_0)} \lambda^{-1/2} \left(1 + O(\lambda^{-1/2})\right) \tag{B}$$

if $x_0 = a$, $S'(x_0) = 0$, $S''(x_0) \neq 0$ (hence, $S''(x_0) < 0$);

$$F(\lambda) = \sqrt{\frac{2\pi}{-S''(x_0)}} f(x_0) e^{\lambda S(x_0)} \lambda^{-1/2} \left(1 + O(\lambda^{-1/2})\right) \tag{C}$$

if $a < x_0 < b$, $S'(x_0) = 0$, $S''(x_0) \neq 0$ (that is, $S''(x_0) < 0$).

Consequently, in accordance with formula (C), as $n \to \infty$ we have

$$\int_{-r}^{r} \left(1 - (x/r)^2\right) e^{\frac{n-3}{2}} dx = \int_{-r}^{r} e^{\frac{n-3}{2} \log\left(1-(x/r)^2\right)} dx \sim r\sqrt{\frac{2\pi}{n}}, \tag{2.5}$$

and, in accordance with formula (A), as $n \to \infty$ and $\delta > 0$ we have

$$\int_{\delta r}^{r} \left(1 - (x/r)^2\right)^{\frac{n-3}{2}} dx = \int_{\delta r}^{r} e^{\frac{n-3}{2} \log\left(1-(x/r)^2\right)} dx \sim r \frac{1}{n\delta} \left(1 - \delta^2\right)^{\frac{n-1}{2}}. \tag{2.6}$$

Therefore no matter how small $\delta > 0$ is, as $n \to \infty$ we have

$$P_n[\delta r, r] \sim \frac{1}{\delta\sqrt{2\pi n}} \left(1 - \delta^2\right)^{\frac{n-1}{2}} \sim \frac{1}{\delta\sqrt{2\pi n}} e^{-\frac{1}{2}\delta^2 n} \to 0. \tag{2.7}$$

This means that the overwhelming part of the area of a multidimensional sphere S^{n-1} is concentrated in a small zone surrounding the equator.

The above circumstance explains the apparently paradoxical fact that if one chooses at random two unit vectors in a space $\mathbb{R}^n$ of large dimension n, then with a very high probability these vectors will turn out to be almost orthogonal (for example, the probability that their scalar product will differ appreciably from zero rapidly decreases as the size of this difference increases). Let us explain this more precisely. All the directions in the space $\mathbb{R}^n$ can be considered to be on an equal footing. The vectors of the pair are chosen randomly and independently. If one of them is chosen, then the other vector will, with high probability, turn out to be in a neighbourhood of the equator, which is orthogonal to the first vector; and this is what formula 2.7 says. This enables us to give an estimate of the deviation from orthogonality.

2.1.4 Isoperimetric inequality and almost constancy of a function on a sphere of very large dimension

We point out and explain one further no less paradoxical and subtle fact relating to multidimensionality.

Let S^m be the unit sphere in a Euclidean space $\mathbb{R}^{m+1}$ of very large dimension $m + 1$. Suppose that on the sphere we are given a fairly regular real function (for example, from some fixed Lipschitz class). We take at random and independently of each other a pair of points and calculate the values of our function at them. With high probability these values will be almost the same and close to some number M_f.

[This (for the moment, hypothetical) number M_f is called the *median value of the function* or the *median of the function*. It is also called the *median value of the function in the sense of Levy*. The motivation for the terminology will soon be explained along with a more precise definition of the number M_f.]

We now explain this phenomenon, but first we introduce some notation and stipulations.

We shall agree that by the distance between two points on the sphere S^m we shall mean its geodesic distance ϱ. We denote by A_δ the δ-neighbourhood in $S^m$of the set $A \subset S^m$. We normalize the standard measure of the sphere by replacing it by the uniformly distributed probability measure μ, that is, $\mu(S^m) = 1$.

We have the following assertions (the proofs of which can be found in the book [3a]).

For any $0 < a < 1$ *and* $\delta > 0$ *there exists* $\min\{\mu(A_\delta) : A \subset S^m, \mu(A) = a\}$ *and it is attained on the spherical cap* A° *of measure* a.

Here A° is defined to be $B(r)$, where $B(r) = B(x,r) = \{x \in S^m : \varrho(x_0, x) < r\}$ and $\mu(B(r)) = a$. For $a = 1/2$, that is, when A° is a hemisphere, we obtain the corollary:

If the subset $A \subset S^{n+1}$ *is such that* $\mu(A) \geq 1/2$, *then* $\mu(A_\delta) \geq 1 - \sqrt{\pi/8}e^{-\delta^2 n/2}$. (Here as $n \to \infty$ $\sqrt{\pi/8}$ can be replaced by $1/2$.)

We denote by M_f the number for which $\mu\{x \in S^m : f(x) \leq M_f\} \geq 1/2$ and $\mu\{x \in S^m : f(x) \geq M_f\} \geq 1/2$. It is also called the *median* or *median value of the function* $f : S^m \to \mathbb{R}$ *in the sense of Levy*. (If the M_f-level of the function f on the sphere has zero measure, then the measure of each of the above two sets will be exactly one half the μ-area of the sphere.)

Levy's Lemma [2] (which immediately follows from the above assertion and its corollary) states the following:

If $f \in C(S^{n+1})$ *and* $A = \{x \in S^{n+1} : f(x) = M_f\}$, *then*

$$\mu(A_\delta) \geq 1 - \sqrt{\pi/2}e^{-\delta^2 n/2}.$$

Now let $\omega_f(\delta) = \sup\{|f(x) - f(y)| : \varrho(x,y) \leq \delta\}$ be the *modulus of continuity of the function* f. The values of the function f on the set A_δ are close to M_f. More precisely, if $\omega_f(\delta) \leq \varepsilon$, then $|f(x) - M_f| \leq \varepsilon$ on A_δ. Thus, Levy's lemma shows that "good" functions really are almost constant on almost the whole domain of definition S^m when its dimension is very large.

This *phenomenon of concentration of measure* close to some value of a function occurs not only for the sphere (see [3a]); although, fairly obviously we do not have this phenomenon for all spaces, functions and measures.

In the book [3a] it is used, for example, for obtaining almost Euclidean subspaces in normed linear spaces of large dimension. (As well as various versions of the proofs of the principle of concentration itself, the book [3a] contains a geometric appendix written by Gromov, including the isoperimetric inquality. This book also contains a bibliography with an indication of the original sources, for example, the book [2] by Levy, where the principle of concentration is stated in explicit form. This phenomenon had already been noted in a certain form by Poincaré in his lectures on probability theory [1]. There is a recent bibliography in the papers [3b], [3c] and [3d].

It is interesting, however, to consider this concentration phenomenon, for example, from the point of view of ergodic theorems in thermodynamic passage to the limit, as well as from the point of view of a geometric interpretation of the law of large numbers (already non-linear) and limiting distributions of probability theory. Possible physical applications are also very attractive (see, for example, the paper [13]).

2.2 Some remarks

2.2.1 *The various averages*

What is the interrelation between the standard average $\bar{f}$ of a function and its median value in the sense of Levy (that is, the quantity M_f introduced above)?

By going over if necessary to the function $f - M_f$ we can assume without loss of generality that for the function f the median value M_f is equal to zero. We denote by $|S^n|$ the (standard Euclidean) area of the n-sphere S^n and let T be the upper bound of the values of the function $|f|$ on this sphere (that is, the maximum of $|f|$ in the case of a continuous function).

Let ε be a small positive number. The calculations

$$
\begin{aligned}
|\bar{f}| &\leq \frac{1}{|S^n|}\left(\int_{|f(x)|\leq\varepsilon T}|f|(x)dx + \int_{|f(x)|>\varepsilon T}|f|(x)dx\right) \\
&\leq \varepsilon T + T\frac{1}{|S^n|}\int_{|f(x)|>\varepsilon T}dx = \left(\varepsilon + \frac{1}{|S^n|}\int_{|f(x)|>\varepsilon T}dx\right)T
\end{aligned}
$$

show that the value of $|\bar{f}|$ is small by comparison with T if the integral in the round brackets is small by comparison with $|S^n|$. In other words, this is so if the area of the region $D_{\varepsilon U} \subset S^n$, where $|f(x)| > \varepsilon T$, is small by comparison with the area of the whole sphere. Now by hypothesis $M_f = 0$, therefore the region $D_{\varepsilon T}$ lies outside some δ-neighbourhood of the median level of the function, around which, as we have shown, the overwhelming part of the entire surface of the sphere is concentrated, of course provided that this neighbourhood itself is not too small. The quantity δ, which characterizes the relative width of this neighbourhood, also just depends on the modulus of continuity of f, which relates the quantities ε and δ.

If we consider functions with a fixed Lipschitz constant L on a sphere of radius r, then for them we have $T \asymp Lr$, $\delta \asymp L^{-1}\varepsilon$ and for $n \gg 1$ with small relative error we have, of course, the equality $\bar{f} = M_f$.

However, one must also show how L increases as n increases; for example (which is typical for thermodynamics), the so-called *summation functions* of the form $f(x_1) + ... + f(x_n)$ (sum of energies of particles, and so on). Already for the simplest linear function $x_1 + ... + x_n$ we have $L = \sqrt{n}$ (for the proof it suffices to transfer from the origin to the point $(1,...,1) \in \mathbb{R}^n$).

In cases of interest for thermodynamics, as we have already remarked, it is natural to assume that the radius of the sphere $S^n(r)$ increases with n and $r \asymp \sqrt{n}$. It is therefore natural to assume also that the range of values of the summation function on such a sphere will be of the order $L \cdot r = \sqrt{n} \cdot \sqrt{n} = n$. In this case one can also establish the validity of this form of the law of large numbers (the phenomenon of concentration described above).

In conclusion we note that in $\mathbb{R}^n$ the standard unit of volume of the n-dimensional "cube" becomes larger and larger as n increases: the diameter of the n-dimensional "cube" is equal to $\sqrt{n}$. A sphere inscribed in such a "cube" takes up a negligible part of the volume of the cube if $n \gg 1$.

It is also helpful to note (and this is used in coding theory; see Ch. 3) that if one takes two balls of the same radius and places them so that the distance between the centres is equal to the radius, then the balls will intersect; however, if $n \gg 1$, then the volume of this region of intersection will be negligibly small by comparison with the volume of each ball.

2.2.2 *The multidimensional cube and the principle of concentration*

We consider the standard n-dimensional unit closed interval $I^n \subset \mathbb{R}^n$, which for brevity we shall allow ourselves to call the n-dimensional cube. We shall peel the cube I^n like an orange by removing from it the $\frac{1}{2}\delta$-neighbourhood of the boundary of the cube. What is left is another cube but now with edge-length $r = 1 - \delta$. Its volume r^n is small if $n \gg 1$. We have removed the volume $1 - r^n = 1 - (1 - \delta)^n$, which forms the main part of the whole volume 1 of the unit cube.

If we had n independent random quantities x_i taking values in the unit interval $[0,1]$ and having probability distributions $p_i(x)$ that are bounded away from zero uniformly with respect to i, for example, if all the $p_i(x)$ were the same, then as n increases the overwhelming part of the random points $(x_1,...,x_n) \in I^n$ would turn out to be in close proximity to the boundary of the cube.

This kind of principle of concentration, formulated appropriately, is of course valid for regions of general type in the space $\mathbb{R}^n$ when $n \gg 1$.

We saw in the example of a sphere that typical values of a function lie close to values which we called median values. Here a large role was played by the geometry of the sphere and the uniformity of the distribution of measure. For regions obtained by a small continuous perturbation of the sphere the principle is still valid, but for large deformations of the sphere or of the measure the principle is no longer valid in its previous form.

Is there a reasonable generalization in which the growth of the dimension is in the final event the overriding factor? (For example, any sufficiently multidimensional convex body admits an almost spherical section of increasing dimension and this "almost" is better the greater the original dimension is.)

2.2.3 *Principle of concentration, thermodynamics, ergodicity*

All the basic thermodynamic functions (for example, pressure) which, from the point of view of statistical physics, are typical values of functions depending on an enormous number of variables (the phase coordinates of the individual molecules) daily demonstrate to us the laws of large numbers and the principle of concentration.

Furthermore, if we consider the evolution of the state of a thermodynamic system as the motion of a point along an energy-level surface or in the very large dimensional region bounded by that surface, then, by the principle of concentration, for most of the time the point will be in the region of median values of any sufficiently regular function defined on this surface or in the region of space bounded by it.

Suppose that the motion proceeds in a region of the multidimensional space $\mathbb{R}^n$ defined by the condition $H \leq E$ (where H is the Hamiltonian of the system). This region is no longer a ball. The layer between the levels H and $H + \Delta$ has non-uniform thickness $\Delta/|\nabla H|$. This needs to be taken into account when comparing the medians in the region and the medians on the surface. If the integration over the region involves the curvilinear coordinate H as one of the variables, then on the surface the uniform standard measure $d\sigma$ is replaced by $d\sigma/|\nabla H|$.

By Liouville's theorem this is the invariant (Gibbs microcanonical) measure corresponding to the Hamiltonian system. However, the function $|\nabla H|$ can also be considered to be almost constant if the level $H = E$ is similar to a sphere.

2.2.4 *Principle of concentration and limiting distributions*

The principle of concentration is the geometric analogue (possibly also non-linear) of the law of large numbers. In classical probability theory the latter usually relates to a linear combination of random quantities, such as their sum. The limit theorems of probability theory establish the law of large numbers and also the limiting probability distribution laws.

Above, by considering the principle of concentration with the example of the area of a multidimensional sphere $S^n(r) \subset \mathbb{R}^{n+1}$, we obtained at the same time the limiting law of the distribution of this area when the dimension of the sphere and its radius $r = \sqrt{n}$ increase unboundedly. The result obtained here corresponds to the central limit theorem of probability theory.

For the perturbed sphere, as we have just remarked, the principle of concentration apparently still applies. This means that other limit theorems (corresponding to various well-known variants of the central limit theorem) must still exist.

At the same time this can give a certain adequate formulation and substantiation of the general opinion that if one imposes on a large system some global constraint (for example the restriction of its total energy) then, in a certain sense, this predetermines its microscopic probability structure.

Before we turn to the next chapter we point out that various aspects of the questions discussed here are explicitly or implicitly present in many investigations (perhaps formally) relating to various areas of mathematics or its applications. In this connection see, for example, the books and papers [1]–[13].

Chapter 3
Communication in the presence of noise

3.1 Discrete recording of a continuous signal — concretization

3.1.1 *Energy and mean power of a signal*

Recall that the sampling theorem (the Kotel'nikov formula)

$$f(t) = \sum_{-\infty}^{\infty} f\left(\frac{k}{2W}\right) \frac{\sin 2\pi W(t - \frac{k}{2W})}{2\pi W(t - \frac{k}{2W})} \tag{3.1}$$

recovers the signal, which is a function $f \in L_2(\mathbb{R})$ with compactly supported spectrum of frequencies ν not exceeding W Hertz from the set of sample values $f(t_k)$ at the points $t_k = k\Delta$, where $\Delta = \frac{1}{2W}$ is the sampling time interval (Nyquist interval), which depends on W.

The wider the frequency band the more complex the function f can be and the more frequently one needs to take samples in order to adequately encode it discretely and recover it, but then the more information it (that is, the signal) can carry.

The function $\operatorname{sinc} t = \frac{\sin t}{t}$ is basic in the expansion (3.1); this function, as we already know, has constant spectrum equal to 1, on the unit interval of frequencies and is the instrumental function of an ideal low-frequency filter with unit pass-band. Thus the sampling function sinc is realized as the response of such a filter to a unit impulse realized at time $t = 0$.

The corresponding function $e_k(t) = \operatorname{sinc} 2\pi W(t - \frac{k}{2W}) = \frac{\sin 2\pi W(t - \frac{k}{2W})}{2\pi W(t - \frac{k}{2W})}$ has spectrum $\check{e}_k(\nu) = \frac{1}{2\pi W}\exp(-i\frac{\pi}{W}k\nu)$ and frequency band $0 \le \nu \le W$ ($|\nu| \le W$).

One can conclude from the orthogonality of the functions $\check{e}_k$ on the interval $[-W, W]$ (or on any interval of length $2W$) and Parseval's equality for the Fourier transform that the functions e_k themselves are orthogonal

V. Zorich, *Mathematical Analysis of Problems in the Natural Sciences*,
DOI 10.1007/978-3-642-14813-2_6,

in the space $L_2(\mathbb{R})$ and $\|e_k\|^2 = \frac{1}{2W}$. Hence we can infer from the equality $f = \sum_{-\infty}^{\infty} x_k e_k$ that $\|f\|^2 = \frac{1}{2W}\sum_{-\infty}^{\infty} x_k^2$.

In practice the signal f has a certain finite duration T, that is, $f(t) \equiv 0$ outside the interval $0 \le t \le T$. This condition is incompatible with the condition that the spectrum of f be compactly supported. However, one can assume that the values $f(t)$ of the function are small outside the interval $[0,T]$ and the sample values of the function outside this interval are set equal to zero.

Then the equality $f = \sum_{-\infty}^{\infty} x_k e_k$ is replaced by $f(t) = \sum_{k=1}^{2WT} x_k e_k(t)$, where $t \in [0,T]$, $x_k = f(k\Delta)$ and $\Delta = \frac{1}{2W}$. This signal f is written by the vector $x = (x_1,...,x_n) \in \mathbb{R}^n$ of its sample values, where $n = 2WT$

Under the same conditions Parseval's equality

$$\int_{-\infty}^{\infty} f^2(t)dt = \sum_{-\infty}^{\infty} x_k^2 \|e_k\|^2 = \frac{1}{2W}\sum_{-\infty}^{\infty} x_k^2$$

is replaced by the equality

$$\int_0^T f^2(t)dt = \sum_1^n x_k^2 \|e_k\|^2 = \frac{1}{2W}\sum_1^n x_k^2 = \frac{1}{2W}\|x\|^2.$$

Here, to within a specific uniform factor, the integral gives the energy (work) of the signal f (for example, when f is realized as the drop in voltage on a unit resistance). Hence the mean power P of the signal f over the time interval $[0,T]$ is

$$P = \frac{1}{T}\int_0^T f^2(t)dt = \frac{1}{2WT}\|x\|^2 = \frac{1}{n}\|x\|^2.$$

Thus, $\|x\|^2 = nP = 2WTP$ and P can be interpreted as the mean power required at one coordinate of the vector x, that is, one sample value of the signal f.

Thus, signals of duration T with compactly supported spectrum in a frequency band W whose mean power is at most P in the vector representation $x = (x_1,...,x_n)$ turn out to be located in a ball $B(0,r) = B(r) \subset \mathbb{R}^n$ of radius $r = \sqrt{2WTP} = \sqrt{nP}$ with centre at the origin of the Euclidean space $\mathbb{R}^n$ of dimension $n = 2WT$.

3.1.2 *Quantization by levels*

The measurement of the sample value of a signal f is performed from a certain threshold (limiting) precision ε. If the amplitude of any signal to be transmitted is not greater than A (that is, $|f|(t) \le A$ for $t \in [0,T]$), then, by endowing the interval $[-A,A]$ with a uniform network of points (levels) with mesh size ε, for $f(t)$ we can take the point of this network nearest to $f(t)$. The values of $f(t)$ turn out to be quantized by levels, the number of which is $\alpha = \frac{2A}{\varepsilon}$ (we take α to be an integer greater than 1). The word $x = (x_1,...,x_n)$ corresponding to the signal f, which consists of n letters x_k will be written in an alphabet having α different characters. In all there are α^n such different words x. If $n = 2WT$ and W and T are large numbers, then α^n is enormous.

3.1.3 Ideal multilevel communication channel

Under these conditions after time T one can distinguish $M = \alpha^{2WT}$ (and not more) different signals $f \sim x = (x_1,...,x_n)$, that is, one can determine one definite signal-word-message out of the M possible ones.

The binary notation $x^0 = (x_1^0,\ldots,x_m^0)$, which distinguishes M objects, requires $m = \log_2 M$ symbols 0,1 (we take m to be an integer). The information about the next coordinate of the binary vector (if the coordinates are on an equal footing and their possible values 0, 1 are equally likely) is taken as the elementary unit of information and is called the *bit*. If we could without any error receive and transmit vectors (words) encoding our M messages, then in time T we could distinguish M objects (signals, messages). The speed of transmitting information (on the choice of one of the M possible objects) along such an ideal communication channel (and with such an encoding) measured in bits per second would be equal to $\frac{1}{T}\log_2 M = 2W\log_2 \alpha$.

3.1.4 Noise (white noise)

We now work with vectors $x = (x_1,...,x_n) \in \mathbb{R}^n$. Here $n = 2WT \gg 1$ and we know that $\|x\|^2 = 2WTP = nP$, where P is the mean power at one coordinate of the vector x , that is, the mean power of the signal f corresponding to x.

Suppose (and this indeed usually happens) that there is noise in the communication channel. It gives rise to a noise vector $\xi = (\xi_1,...,\xi_n) \in \mathbb{R}^n$ and at the receiving end of the communication channel, instead of the vector x, the displaced vector $x + \xi$ is received. Thus around each point $x \in \mathbb{R}^n$ there occurs a region of uncertainty $U(x)$ at points of which noise can displace x.

Noise can be of different kinds; accordingly it can have various characteristics. We shall assume that our noise is random, is independent of x and is white (thermal) noise, that is, the vector $\xi \in \mathbb{R}^n$ is random and its coordinates are independent random quantities identically distributed in accordance with the normal Gaussian law (with zero mathematical expectation and variance σ^2). Let N be the mean (at the sample value) power of the noise. Then $\|\xi\|^2 = nN = 2WTN$ (N comes from the word "noise" and P

from the word "power") and $\sqrt{N} = \sigma$ is the standard deviation of the random value of each coordinate of the vector ξ. As before, we assume that $2WT = n \gg 1$.

3.2 Transmission capacity of a communication channel with noise

3.2.1 *Rough estimate of the transmission capacity of a communication channel with noise*

The combined mean power of the signal and the noise is at most $P + N$, therefore the coordinates of the vector $x + \xi$ in modulus should on average not exceed $\sqrt{P + N}$ and it should lie inside a ball of radius $\sqrt{n(P + N)}$.

Since the expected displacement of each of the coordinates of the vector (of information) x under the influence of the noise (white noise) is of order $\sigma = \sqrt{N}$, the number of well distinguishable values of each coordinate at the receiving end is proportional to $\sqrt{\frac{P+N}{N}}$. The coefficient of proportionality k depends on how one interprets the phrase "well distinguishable". If one is required to improve the resolution, then k needs to be made smaller.

In time T there are $n = 2WT$ independent values (of the samples) of the coordinates, therefore the total number K of distinguishable signals will be $\left(k\sqrt{\frac{P+N}{N}}\right)^{2WT}$. Hence, the number $\log_2 K$ of bits that can be transmitted in time T will be $WT \log_2 k^2 \frac{P+N}{N}$. This means that the speed of transmission will be $W \log_2 k^2 \frac{P+N}{N}$ bits per second.

3.2.2 *Geometry of signal and noise*

We now recall the following. On the vector $x \in \mathbb{R}^n$ an obstacle $\xi \in \mathbb{R}^n$ is imposed in the form of white noise. This means that we are given a vector $x \in \mathbb{R}^n$ and a random vector $\xi \in \mathbb{R}^n$ that is independent of x and has uniform distribution along the directions of $\mathbb{R}^n$. The dimension n of the space $\mathbb{R}^n$ is enormous. Then it follows from the principle of concentration, which we discussed in Chapter 2, that with negligibly small probability of error the vector ξ will be almost orthogonal to the vector x (that is, the scalar product and the correlation of the vectors x and ξ should be considered to be zero).

We add to this that, in view of the concentration of the main part of the volume of a multidimensional ball in a small neighbourhood of its boundary sphere, we can suppose that if a random point lies in such a ball, then it is most likely to be situated almost on its boundary. Thus in our situation, when $n = 2WT \gg 1$, we are justified in supposing that $\|x\|^2 = nP$, $\|\xi\|^2 = nN$, $\|x+\xi\|^2 = n(P+N)$.

The regions of uncertainty $U(x)$ arising at the receiving end around each point $x \in \mathbb{R}^n$ as a result of the influence of noise can, in our case, be considered to be balls $B(x,r)$ of radius $r = \sqrt{n\sigma} = \sqrt{nN}$.

Under these conditions how many distinguishable signals are there in the ball $B(0,\sqrt{nP})$? Clearly, not more than the ratio of the volume of the ball $B(0,\sqrt{n(P+N)})$ to the volume of a ball of radius $\sqrt{nN}$. Thus, we have the following upper estimate for the number M of distinguishable signals:

$$M \leq \left(\sqrt{\frac{P+N}{N}}\right)^{2WT} = \left(\frac{P+N}{N}\right)^{WT}, \tag{3.2}$$

which means that we have the following estimate for the speed C of transmission of information:

$$C = \frac{\log_2 M}{T} \leq W \log_2 \frac{P+N}{N} = W \log_2 \left(1 + \frac{P}{N}\right). \tag{3.3}$$

Here it is worth pausing and making some observations. If one tries to pack as many balls of radius $\sqrt{nN}$ as possible in a ball of radius $\sqrt{n(P+N)}$ under the condition (as a presupposition) that the inserted balls are, as it were, rigid and non-intersecting, but can abut one another, then for $n = 2WT \gg 1$ the number of such balls will be catastrophically small by comparison with the ratio of volumes indicated above. Shannon's theorem (see below), whose proof we are now ready for, states that, nevertheless, for sufficiently large times T we can get the speed of transmission to be as near as we like to the upper-estimate quantity C indicated above and also having an arbitrarily small probability of an error when transmitting the message.

Just the possibility of making some errors, although as rare as one pleases, eliminates the condition that the inserted balls do not intersect. If the dimension of the space is large, then, as we noted in Chapter 2, the centres of the balls can get close to one another, the balls will intersect, but the relative volume of their intersection can be very small even when the centres of balls of the same radius are at a distance equal to the length of the radius. As the centres approach one another, the number of inserted balls increases, but then also does the probability of an error when decoding the received signal.

The calculation of the interaction of the above circumstances forms the geometric basis of Shannon's theorem.

3.2.3 Shannon's theorem

Theorem. *Let P be the mean power of the transmitter and suppose that we have white noise with power N in the frequency band W. Then by applying a sufficiently complex system of coding it is possible to transmit binary digits with speed*

$$C = W \log_2 \frac{P+N}{N}$$

with arbitrarily small frequency of errors. No method of coding can be transmitted with greater average speed and with arbitrarily small frequency of errors.

We take M to be the number on the left-hand side of the estimate (3.2). This is a large number and we assume that it is an integer (by ignoring its fractional part). In the ball $B(0, \sqrt{nP}) \subset \mathbb{R}^n$, where the vectors x (words, signals) to be transmitted are, we choose M points at random. Here by "at random" we mean that the points are chosen independently at random and the probability that a point hits some region is proportional to the volume of that region, that is, it is equal to the ratio of the volume of that region to the volume of the whole ball $B(0, \sqrt{nP})$. (If the random choice of M balls is repeated many times, then, as a rule, the points will be distributed in the above fashion.) We have $n = 2WT$ and $M = \left(\frac{P+N}{N}\right)^{WT}$, therefore one point will be arriving in a volume of size $\frac{1}{M}|B|$, where $|B|$ is the volume of the whole ball $B(0\sqrt{nP})$. Hence the probability that one of our M points will hit this same region is equal to $\frac{1}{M} = \left(\frac{N}{P+N}\right)^{WT}$. As $T \to +\infty$ this probability of course tends to zero independently of the ratio of the positive quantities P and N.

If our M points are chosen randomly, then, assuming that $n = 2WT \gg 1$ and the volume of the ball $B(0, \sqrt{nP})$ is concentrated near its boundary sphere, one can with negligible relative error (which is smaller the larger T is) assume that all the chosen points will be in an arbitrarily small neighbourhood of the boundary sphere.

We recall further that the noise vector ξ, as we showed earlier, when $n \gg 1$ is orthogonal to the signal vector x with probability arbitrarily close to 1. Thus, for $n = 2WT \gg 1$ (that is, as $T \to +\infty$) we have $\|x\|^2 = nP$, $\|\xi\|^2 = nN$, $\|x + \xi\|^2 = n(P+N)$.[1]

We now proceed with the concluding arguments and obtain a concrete estimate. Suppose that we have made a typical random selection of M points in the ball $B(, \sqrt{nP})$. Suppose that they correspond to M different messages that we intend to send along a communication channel. Such a choice of points in the ball together with their corresponding messages means a cer-

[1] If P and N are interpreted as the variances D_x, D_ξ of the signal and noise, then here we have the classical probability-theoretic relation for the variance of the sum of independent random quantities: $D^2_{x+\xi} = {D_x}^2 + {D_\xi}^2$.

tain encoding of the messages intended for transmission. We coordinate in advance the chosen code with the receiving device. If there were no noise, then the receiver, having received a signal x without corruptions, would uniquely decipher it into the message corresponding to it in accordance with the agreed code.

In the presence of noise in the channel, instead of x one will get $x + \xi$ at the receiving device. The receiver looks for the point in the ball $B(0, \sqrt{nP})$ among the points of the fixed code that is nearest to $x + \xi$ and takes it to be the transmitted signal. Here there is the possibility of error, that is, it is possible not to read the message that was sent. However, this is possible only if there is one of the M points of the code apart from x in the ($\|\xi\| = \sqrt{nN}$)-neighbourhood of $x + \xi$.

We find an upper estimate of the probability of such an event. For this we estimate the volume of the intersection of a $(\sqrt{nN})$-neighbourhood of a point $x + \xi$ with the ball $B(0, \sqrt{nP})$. Since $\|x + \xi\|^2 = n(P + N)$, this is a simple geometrical problem. Consider a two-dimensional plane passing through the origin 0 and the points $a = x$ and $b = x + \xi$. The triangle $0ab$ is right-angled with right angle at the vertex a and with side lengths of the legs $|0a| = \sqrt{nP}$ and $|ab| = \sqrt{nN}$ and hypotenuse $|0b| = \sqrt{n(P + N)}$. By calculating its area by two methods we easily find the length h of the perpendicular drawn from the vertex a to the hypotenuse: $h = \sqrt{n\frac{PN}{P+N}}$. If we now take a ball of radius h and centre at the base of this perpendicular, then clearly it covers the entire region (of interest to us) of intersection of the ball $B(0, \sqrt{nP})$ and the $(\sqrt{nN})$- neighbourhood of the point $b = x + \xi$. Hence, the probability that, along with x, which lies on the boundary of this region, there will also be one of the M code points is less than the ratio of the volume of a sphere of radius h to the volume of a sphere of radius $\sqrt{nP}$. Thus, this probability is less than $\left(\frac{N}{P+N}\right)^n = \left(\frac{N}{P+N}\right)^{WT}$, and this tends to zero as $T \to +\infty$.

Thus, for sufficiently large values of T, with arbitrarily small probability of error in such a communication channel one can distinguish one of $M = \left(\frac{P+N}{N}\right)^{WT}$ different objects; more precisely, one can identify one of the M possible different messages in time T. In terms of binary units this is $\log_2 M$ bits of information in time T. Hence we can indeed achieve the speed of transmission arbirtrarily close to the upper bound estimation indicated in inequality (3.3).

This completes the proof of the theorem.

3.3 Discussion of Shannon's theorem, examples and supplementary remarks

3.3.1 *Shannon's commentary*

The best brief commentary shedding light on certain aspects of the theorem which at first reading are seldom noticed is the following commentary by Shannon himself [2]:

"We shall call a system that transmits with speed C and without errors an ideal system. Such a system cannot be realized by any finite encoding process but can be approximated as closely as one pleases. The following happens when the approximation approaches the ideal: 1. The speed [2] of transmission of binary numbers approximates to $C = W\log_2(1 + P/N)$. 2. The frequency of errors approximates to zero. 3.The signal to be transmitted approximates to white noise in its statistical properties. Roughly speaking this is true because the coding points are randomly distributed inside a ball of radius $\sqrt{2WTP}$. 4. The threshold effect becomes sharp. If the noise exceeds the value for which the system was constructed then the frequency of errors increases very rapidly. 5. The required delays in the transmitter and receiver increase unboundedly. Of course in a broad-band system a delay of one millisecond can already be considered to be infinite."

Here perhaps an explanation is required only for the first sentence in item 5, which at the same time also explains the real meaning of the quantity C featuring in the theorem as the speed of transmission. To write down in bits each of M different objects requires $\log_2 M$ bits. An individual message of the M possible ones is sent or arrives only after what is transmitted (respectively, received) takes the whole of its binary code of length $\log_2 M$. For this time $2T$ is required too, which also implies a delay of the same message, while at the same time, as $T \to +\infty$, the mean speed of transmission of the bits (bits per second) in fact approximates to the upper limit indicated in the theorem.

Later on we shall give some examples which possibly will explain certain aspects of the range of questions touched upon here.

3.3.2 *Weak signal in a large amount of noise*

It is clear from the construction of an optimal code (and this is explicitly pointed out by Shannon in item 3 of the above quotation) that in its statistical properties such a code is similar to white noise. This means that establishing

[2] *Author's remark.* Like the Meshcherskiĭ–Tsiolkovskiĭ formula for the speed of a rocket!

(or discovering) contact with a very intelligent extraterrestrial civilization sending us signals indistinguishable from noise is fairly difficult.

But let us consider the following situation. Arriving towards us is a weak periodic signal in the background of a large amount of noise that is, however, random. For example, in the channel there is a large amount of white noise. Is it possible to separate the useful signal f from the noise? Suppose that we know or could know the period T of the signal f. Let us listen to and record the signal with the noise $n \gg 1$ times. We then reproduce all these signals synchronously, that is, we put them together. Then the random noise will itself be dampened, while the signal will be strengthened. This means that sometimes one can combat the noise by actually making use of it.

3.3.3 *Redundancy of language*

In a communication channel we very often cope with noise in a similar way, which the science calls complicating the code or its redundancy.

In item 4 of his commentary Shannon noted the sharp threshold effect of the optimal code. We shall return to this a little later, but meanwhile we give some explanation with an example.

If in a telephone conversation you are dictating something to the person at the other end and there is some word that he could not make out or does not know, then you start to repeat or communicate letter by letter, while the letters are communicated by pronouncing entire words such as Anna, Maria, Booby, Aristotle, and so on.

You fight against the noise by encoding A, M, B, etc., with a code that is certainly redundant. An optimally economical code is, of course, splendid, but also dangerous, as is every maximum of potential possibilities — it is unstable. Any spoken language, as we can easily observe, is redundant (roughly by 50%) but, on the other hand, good for everyday intercourse.

3.3.4 *Precise measurements in a crude piece of apparatus*

How does one measure the thickness of a sheet of paper on an apparatus where you have measured your height only to within 0,5cm? Recall the example given above of a weak signal in a communication channel with a large amount of noise. If there is the possibility of taking some packet of this paper, and by adding them one can find, for example, that one thousand sheets of paper have a thickness of 20cm with absolute error within the limits of 0,5cm, then, assuming that the sheets are all roughly the same, we find that the thickness of one sheet is 0,2mm with possible error in the limits of 0,005mm.

The idea of the above examples can be extended to create accurate constructs (devices, apparatuses) from inaccurate elements.

3.3.5 *Shannon–Fano code*

Hidden in the numerous details of the proofs the probabilistic structure of an optimal code in Shannon's theorem is clearly distinguishable in all its detail in the following naked idea of an optimal code, called the Shannon–Fano code. For economy of space and words we consider a simple demonstrative example, the possible generalization of which is obvious.

We have an alphabet of four letters from which words are formed. Along a communication channel the bits 0 and 1 can be transmitted with the same speed and accuracy. One can encode the letters of our alphabet as follows: (0,0), (0,1), (1,0), (1,1). After this one can transmit text in these letters. Meanwhile we forget about the noise and concern ourselves with the economy of the code, which influences the speed of the transmission of information.

We suppose that a statistical analysis of the language establishes that the four letters of the alphabet are encountered with different frequencies; for example, suppose that their probabilities are $\frac{1}{2}, \frac{1}{4}, \frac{1}{8}, \frac{1}{8}$, respectively. Then it is most reasonable to proceed as follows. First we divide the letters into two equiprobable groups (here it is the first letter and all the remaining letters), which we distinguish by the symbols 0 and 1, respectively. We then repeat the same procedure with each of the groups and their subgroups as long as the subgroups do not reduce to a single element. And this is the idea of the Shannon–Fano code. In our case the code looks like this: (0), (1,0), (1,1,0), (1,1,1). We compare the above two codes on a sufficiently long text of T letters. In the first case we need to send $2T$ bits. In the second case it is $(\frac{1}{2}1 + \frac{1}{4}2 + \frac{1}{8}3 + \frac{1}{8}3)T = \frac{7}{4}T$ bits. Moreover, even without punctuation signs in the optimal code one can recover the sequence of letters (10011000111110 — decipher that). But one only has to make an error in the transmitter or receiver of just one bit and the text becomes unreadable.

3.3.6 *Statistical characteristics of an optimal code*

The above example demonstrating the idea behind the Shannon–Fano code shows that an optimal code tends to distribute the information uniformly in terms of the symbols transmitted. This can be achieved for the transmission of long messages provided that their statistical processing is dealt with beforehand. The dangers of an optimal code are now also clear.

We note further that Shannon's theorem relates to noise assumed to be white noise. Noise can be of various kinds, both random and deterministic. Moreover, even random noise can have various statistical characteristics. Obviously in a specific situation one needs to act in a specific manner. The general theory gives indications on the reasonable order of such actions but does not solve the whole problem in one go.

3.3.7 *Encoding and decoding — ε-entropy and δ-capacity*

Earlier we already mentioned the quantization of a continuous signal into a discrete number of levels. The standard procedure enabling one to go over to a discrete finite description of any compact subset of a metric space consists in the construction of a finite ε-net, that is, a finite collection of points such that any point of the compact set can to within an ε-shift be approximated (replaced) by one of the points of this net. The quantity ε characterizes the allowable accuracy of the approximation, or the allowable error. If the devices being used are not capable of distinguishing objects in scales less than ε, then, without special need, it makes no sense to bother with all the points of the compact set; it suffices to replace the compact set by some ε-net of it. The ε-net itself can be considered to be a discrete code describing the compact set to within an accuracy of ε.

Of course it is desirable to have an ε-net with the greatest possible economy, that is, containing the fewest possible number of points. When ε tends to zero the number N_ε of points in such an ε-net of greatest economy increases without bound in general. Its rate of increase is related to the specific character of the compact set and the metric space.

Kolmogorov called the quantity $\log_2 N_\varepsilon$ the *ε-entropy* of the compact set. If, for example, one takes the unit cube I^n or any bounded region of Euclidean space $\mathbb{R}^n$, then, as is easily verified, the limit of the ratio of $\log N_\varepsilon$ to $\log \frac{1}{\varepsilon}$ as ε tends to 0 gives the dimension n of the space.

Incidentally one can exploit the above circumstance to redefine dimension and thus have the possibility of talking about dimensions that are not necessarily integers.

Another more interesting example of the use of ε-entropy which is worth recalling relates to Hilbert's thirteenth problem [3]. Roughly speaking, the question can be restated in the following eye-catching form: do functions of several variables exist? More precisely: can every function of several variables be assembled from functions of a smaller number of variables, that is, can it be represented as a superposition (composite) of finitely many such functions?

A.N. Kolmogorov and V.I. Arnold proved that every continuous function of several variables can be represented as a superposition of continuous functions of one and two variables; here, as Kolmogorov noted, for the function of two variables it suffices to have only the function $x + y$ (see [5], [6a], [6b]).

But even before that A.G. Vitushkin had proved that not every smooth function of several variables is the superposition of functions of a smaller number of variables and enjoying the same amount of smoothness; see [4]. To state this precisely, after Vitushkin we consider a number $v = \frac{n}{p}$ — the ratio of the number of variables to the order of its highest-order continuous derivatives. It serves as the index of complexity of the function in the sense of Vitushkin. As always, we denote by $C_n^{(p)}$ the class of p-smooth functions of n variables defined on the unit n-dimensional cube $I^n \subset \mathbb{R}^n$. Let $k < n$. The question is: when can any function of class $C_n^{(p)}$ be represented as a superposition of functions of class $C_k^{(q)}$? Vitushkin showed that this is possible only if $(\frac{n}{p} = v) \leq (\tilde{v} = \frac{k}{q})$.

Vitushkin's proof used, in particular, Oleĭnik's estimates of the Betti numbers of the algebraic manifolds obtained in connection with investigations by her and Petrovskiĭ of Hilbert's 16th problem on the number and positioning of ovals of a real algebraic curve.

Kolmogorov gave another direct and, seemingly the most natural, explanation (proof) of Vitushkin's result precisely in connection with information and entropy [7a], [7b].

The spaces $C_n^{(p)}$ and $C_k^{(q)}$ are infinite-dimensional but, as Kolmogorov showed, if in these spaces one takes the compact sets consisting of all functions whose derivatives are bounded by some fixed constant, then as $\varepsilon \to 0$ their ε-entropy will increase as $(\frac{1}{\varepsilon})^{\frac{n}{p}}$ and $(\frac{1}{\varepsilon})^{\frac{k}{q}}$, respectively. If all the functions in $C_n^{(p)}$ can be represented as a superposition of finitely many functions of class $C_k^{(q)}$, then $\left(\frac{1}{\varepsilon}\right)^{n/p} = O\left(\left(\frac{1}{\varepsilon}\right)^{k/q}\right)$. Thus, the inequality $(\frac{n}{p} = v) \leq (\tilde{v} = \frac{k}{q})$ must hold.

As regards Hilbert's problem itself it is worth noting nevertheless that within the framework of algebraic functions (which possibly Hilbert himself also had in mind when speaking about his thirteenth problem on the representation of solutions of a seventh-degree algebraic equation) that the problem is still open. In this connection see the sources [3], [7a] and [7b].

We do not intend here to delve too deeply into these questions and we have only mentioned an example of another non-trivial use of the notion of discrete code and ε-entropy.

Again we return briefly to discrete encoding and add a few words about decoding. An economical ε-net can serve as an economical discrete code of an object (compact metric space) describing it to within an accuracy of ε. Suppose that a specimen of such a code is at both ends of a communication channel. If there is no error in the transmission of the message, then at the receiving end a signal is obtained about that point of the ε-net that was selected at the transmitting end. But if for some or other reason errors in the communication channel are possible, and the transmitted point can be interpreted within the limits of its δ-neighbourhood, then clearly errors are possible during the decoding process; and we already spoke about that earlier. If we are going to exclude the possibility of errors completely, then we have to refrain from using an economical code in the form of an ε-net. By contrast we now have to seek a maximal collection n_δ points of the compact set separated from each other by a distance of at least 2δ. Only such a code (it clearly will be a 2δ-net) can, under the above conditions, guarantee error-free transmission.

While the quantity $\log_2 N_\varepsilon$ is called the *ε-entropy*, as we already know, the quantity $\log_2 n_\delta$ is called the *δ-capacity*.

Calculation of the ε-enropy and δ-capacity of various classes of functions can be found in [8], [9]. Some further information relating to signal processing can be found in [9]–[14].

3.4 Mathematical model of a channel with noise

3.4.1 Simplest model and formulating the problem

As usual, for economy of everything we consider to begin with the simplest model, which however already contains almost everything of most value for our needs for the moment and can easily be generalized if one wishes.

In a communication channel the transmitter sends to the receiver the symbols 0 and 1. The noise results in the possibility that from time to time the receiver deciphers the sent symbol 0 as 1, and 1 as 0. Let p be the probability of a correct passage of the transmitted symbol.

Sent along the channel are messages (text, words) consisting of successive letters (symbols 0, 1 of our two-letter alphabet). We suppose that the channel acts on each letter of the word independently, that is, it is a *channel without memory*. What is the transmission capacity of such a channel?

So that is our problem. Intuitively it is clear that it is a reasonable problem. At the same time it is clear that for its answer it needs to be made clear what exactly one has in mind.[3]

Earlier, before the proof of Shannon's theorem, (after Shannon) we had to sort out the meaning and precise content of certain terms and concepts which our intuition allowed us to use. We now implement this task (again after Shannon). Of course our earlier attempts will considerably lighten this task. Properly speaking it will largely be an abstract formulation of it.

---

[3] It is recounted that one of the visitors of the celebrated Princeton Institute of Advanced Study was housed in Gödel's study, which was temporarily vacated. On leaving, the visitor left a thank-you note on the table expressing his regret that he had not made a closer acquaintance with Gödel. After a while he received a polite letter from Gödel, who had read the note, which asked him to clarify what exactly he had in mind.

As regards a more general abstract model of a communication channel with noise, it is clear that instead of an alphabet of two letters one can have any finite (but not one-letter) alphabet and have the probability that the ith sent letter is converted to the jth at the receiving end. The matrix (p_{ij}) of conversion probabilities models a communication channel with noise.

If the letters are corrupted with different probabilities, then the speed of transmission of information can depend (and does depend) on the cleverness of the code used for writing the messages to be transmitted. Clearly it is best to use most often those symbols that are least subject to corruption. Furthermore, as we already know by experience of the Shannon–Fano code, it is helpful to take into account the statistical peculiarities of the text of the very message subject to transmission.

Apparently, the transmission capacity of a channel of given matrix (p_{ij}) must simply be the upper bound of the possible speed of transmission with respect to everything that the channel (device) itself does not depend on, for example, the upper bound of all possible encodings of the texts to be transmitted. Clearly different users can use one and the same device with different degrees of efficiency. The capabilities of the device itself must be evaluated under the assumption that it is used with maximum efficiency.

After maximal speed under an optimal code has been achieved there can clearly emerge new problems. For example, we saw what dangers are hidden behind codes of maximal economy. But let us lay all this to one side. Just now we need to gradually investigate the question of the speed of transmission of information and what in general we mean by the terms *information* and *quantity of information*.

3.4.2 Information and entropy (preliminary considerations)

As we have already remarked, the appearance of the telegraph and wireless communication stimulated the development of the concept of information and its quantitative description.

It would seem that the measure of information can reasonably be considered to be the measure of the change of uncertainty associated with the information received.

In the simplest situation when there are two possibilities on an equal footing, for example, when the random quantity has exactly two equiprobable values 0 and 1 (off, on), the information about its concrete value (state) liquidates the uncertainty. Recall that the measure of such information deals with what is called the *bit* (short for binary digit).

To identify one of the M objects by putting questions to which the answers are only "yes" (1) or "no" (0) requires, as is well known, $\log_2 M$ binary symbols (the repeated-bisection algorithm). Such a system (random quantity) is capable of storing $m = \log_2 M$ bits of information (corresponding to the measure of its uncertainty). More precisely, if all M possible values (states) of the random quantity under consideration are equally likely, then the identification (selection, information on realization) of one of them under the indicated correspondence is equivalent to giving $\log_2 M$ bits, that is a message of $\log_2 M$ bits of information.

We now state this more formally. Let X be an arbitrary discrete random quantity that can take M different values x_i with probabilities p_i, respectively. How does one take into account the probabilities ? What measure of uncertainty (and information) is it reasonable to associate with such a random quantity ?

We write down the result just obtained by us in the following form:

$$m = \log M = M \cdot \frac{1}{M} \log M = \sum \frac{1}{M} \log M = -\sum \frac{1}{M} \log \frac{1}{M},$$

where we treat $\frac{1}{M}$ as the probability of the appearance (realization, selection) of a specific one of these M objects. (Here and in what follows $\log = \log_2$.)

Then, surely, in general we should arrive at the quantity $-\sum_{i=1}^{M} p_i \log p_i$. We now substantiate this assertion. The quantity $H(X) = -\sum p_i \log p_i$ is called the *entropy* of the discrete random quantity X. (By continuity we suppose that $0 \log 0 = 0$.)

Let us experiment with this. If the probability p_i of an event x_i is small, then the information that this very rare event has occurred can be taken to be the very large number $-\log p_i$. On the other hand, if the event is rare, then over a long period of observations it appears with its information in its teeth extremely rarely (a fraction p_i of the whole time of observations). Therefore the information averaged over a large time interval of observations which this event yields (the value x_i of the random quantity X) is equal to $-p_i \log p_i$.

Thus, if $-\log p_i$ is the measure of uncertainty and information associated with the event x_i whose probability is p_i, then $-p_i \log p_i$ is the average statistical quantity of information that the appearance of such an event yields, and then $H(X) = -\sum_{i=1}^{M} p_i \log p_i$ (mathematical expectation of $-\log p_i$) is the average quantity of information that a single event (value) of the random quantity X carries.

Be aware of the fact that we are not interested in what exactly the real event x_i consists in, although for other purposes it may be that this is the most important thing.

We now settle on a precise notation for the statistical character of entropy: for any positive numbers ε, δ there is a number $n_{\varepsilon\delta}$ such that for $n \geq n_{\varepsilon\delta}$ we have the inequality

$$P\{| -\frac{1}{n} \sum_{i=1}^{n} \log p_{x_i} - H(X)| < \delta\} > 1 - \varepsilon, \tag{3.4}$$

where, as usual, P is the probability of the event indicated in the curly brackets, but now the x_i, $i = 1,...,n$, are n independent values of the random quantity X and the p_{x_i} are the probabilities of these values.

How is the entropy related to the encoding?

Consider the message-word-vector $\bar{x} = (x_1,...,x_n)$ formed by n successive independent values of the random quantity X. The probability $p_{\bar{x}}$ of the appearance of the word $\bar{x}$ is equal to $p_{\bar{x}} = p_{x_1} \cdot ... \cdot p_{x_n}$. In view of formula 3.4 for $n \geq n_{\varepsilon\delta}$, with probability $1 - \varepsilon$ we have

$$2^{-n(H(X)+\delta)} \leq p_{\bar{x}} \leq 2^{-n(H(X)-\delta)}. \tag{3.5}$$

The word $\bar{x}$ is called *δ-typical* if these estimates hold for it. Clearly there exist at most $2^{n(H(X)+\delta)}$ such δ-typical words, while if $n \geq n_{\varepsilon\delta}$, then there are at least $(1-\varepsilon)2^{n(H(X)-\delta)}$ of them and the entire set of non-δ-typical words has probability at most ε.

In principle, now we can already use binary sequences of length $n(H(X) + \delta)$ to encode all δ-typical words. Even if all the remaining words are encoded with one symbol, the probability of an error in transmitting words $\bar{x}$ of length n invoking such a code will be less than ε.

On the other hand (and we have already mentioned this effect of the instability of economical codes), any code using in the same situation binary sequences of relatively slightly smaller length $n(H(X) - \delta)$ (for example, $2\delta n$ out of the $n(H(X) + \delta)$ sent symbols were lost in the noise) will have an asymptotically non-vanishing probability of an error, which tends to one as $n \to +\infty$.

Thus the relation between entropy and encoding of information consists, for instance, in the fact that as $n \to +\infty$ an efficient encoding requires $N \sim 2^{nH(X)}$ words and the entropy $H(X)$ can be interpreted as a measure of the quantity of information in bits in the symbol being transmitted, that is, in one value of the random quantity X.

Hence it follows, in particular, that the entropy of the source of the information should not exceed the capacity of the communication channel if we wish adequately and without delays to transmit the information at hand along this communication channel.

3.4.3 *Conditional entropy and information*

We turn step-by-step to the transmission of information along a communication channel. The transmitter sends the message $\bar{x} = (x_1,...,x_n)$ and the receiver receives $\bar{y} = (y_1,...,y_n)$. How does one recover what was sent from what was received? If there are no corruptions, that is, $y_i = x_i$ always, then there is no problem. We therefore assume that the channel is characterized by some matrix (p_{ij}) of probabilities that the transmitted signal x_i will be converted to the received signal y_j.

We put the question another way. What information about the message $\bar{x}$ is contained in the message $\bar{y}$? Or in other words: how is the uncertainty of $\bar{x}$ changed (decreased) when we know $\bar{y}$?

We turn to conditional probabilities and introduce the concept of conditional entropy $H(X|Y)$ of a random quantity X at the input of a communication channel with respect to the random quantity Y at the output. Then, after Shannon, we consider the quantity

$$I(X;Y) = H(X) - H(X|Y), \tag{3.6}$$

and regard it as the effective *quantity of information* that, on average, is transmitted by one sent signal (value of the random quantity X) in this communication channel.

Hence, the *capacity of the communication channel* is defined as

$$C = \sup_{\{p_x\}} I(X;Y), \tag{3.7}$$

where the supremum is taken over all possible codes, that is, over all possible probability distributions $\{p_x\}$ of the input random quantity X, which has a fixed finite set of values (alphabet).

Thus, we define the conditional entropy $H(X|Y)$ of one random quantity X with respect to another random quantity Y.

Let $\{p_x\}$, $\{p_y\}$ and $\{p_{x,y}\}$, respectively, be the probability distributions of random quantities X, Y and the joint random quantity $Z = (X,Y)$. If the probability of the appearance of the value x_i at the input of the random quantity X is equal to p_{x_i}, and the probability $p(y_j|x_i)$ of conversion of x_i to y_j is given and is equal to p_{ij}, then the probability p_{x_i,y_j} of the combined event $z_{ij} = (x_i, y_j)$ is equal to $p(y_j|x_i)p_{x_i}$, and the total probability p_{y_j} of the appearance at the output of the value y_j of the random quantity Y is equal to $\sum_i p(y_j|x_i)p_{x_i}$.

To ease the notation and without losing any clarity we shall no longer write the extra lower indices. For example, we shall write the standard formula for conditional probability as follows: $p_{x,y} = p(y|x)p_x$ or $p_{x,y} = p(x|y)p_y$, since $p_{x,y} = p_{y,x}$.

First we find the conditional entropy $H(X|Y = y)$ of a random quantity X under the condition that the random quantity Y has the value y. In other words, we now find what the entropy (uncertainty) X becomes under the condition that the random quantity Y has taken the value y.

$$H(X|Y = y) = -\sum_x p(x|y) \log p(x|y) = -\sum_x \frac{p_{x,y}}{p_y} \log \frac{p_{x,y}}{p_y}.$$

We can now find what we are interested in, namely, the conditional entropy $H(X|Y)$ of the random quantity X with respect to the random quantity Y:

$$H(X|Y) = -\sum_y p_y H(X|Y=y) = -\sum_y p_y \sum_x \frac{p_{x,y}}{p_y} \log \frac{p_{x,y}}{p_y} =$$

$$= -\sum_{x,y} p_{x,y} \log p_{x,y} + \sum_y p_y \log p_y = H(X,Y) - H(Y).$$

Here $H(X,Y)$ is the combined entropy of the pair $Z = (X,Y)$ of random quantities X and Y; the probability distribution of the pair is $\{p_{x,y}\}$.

We have found that $H(X|Y) = H(X,Y) - H(Y)$. But since $p_{x,y} = p_{y,x}$ and $p_{x,y} = p(y|x)p_x = p(x|y)p_y = p_{y,x}$, we also have the relations $H(X,Y) = H(Y,X)$ and $H(X,Y) = H(X|Y) + H(Y) = H(Y|X) + H(X)$. Thus,

$$H(X,Y) = H(X|Y) + H(Y) = H(Y|X) + H(X) = H(Y,X). \tag{3.8}$$

Taking into account formula (3.6) (Shannon's definition) for the quantity of information we find that

$$I(X;Y) = H(X) - H(X|Y) = H(X) + H(Y) - H(X,Y). \tag{3.9}$$

Since $H(X,Y) = H(Y,X)$, it follows that

$$I(X;Y) = I(X;Y). \tag{3.10}$$

3.4.4 *Interpretation of loss of information in a channel with noise*

We pause briefly to summarize informally the purport of the concepts introduced above and the interactions that have been uncovered.

The entropy $H(X) = -\sum_x p_x \log p_x$ of a discrete random quantity X is a certain statistical average of its character. If $-\log p_x$ is treated as the measure of uncertainty of the rareness of the event (of the value x of the random quantity X), expressed in bits, and measuring the quantity of information contained in the message about the occurrence of the event x is proportional to this, then $H(X)$ will be the mathematical expectation of this quantity $-\log p_x$.

The entropy is some average measure of the uncertainty of a random quantity X taking on one of its values. Put another way, it is the average measure of information that arrives at one value of the random quantity. Here it is assumed that we obtain a linear series of independent values of the random quantity X and we average the quantity of information obtained from the number of received values of the random quantity. We also tacitly assume that the values are transmitted and dealt with uniformly — one value per unit of time. Therefore in the case when we are dealing with the transmission of information along a communication channel one prefers to treat the entropy of the source as the average quantity of information that the source creates in a unit of time.

If the channel is capable of transmitting this flow of information without corruptions, then everything is fine. If, on the other hand, errors can arise, then we have new problems on our hands. In the example of Shannon's theorem it is clear that in a concrete situation it is necessary to deal with concrete physical parameters (frequency band, signal level, noise level, statistical characteristics of the noise, and so on). Properly taking into account and handling these parameters is an important problem in its own right.

We considered some abstract model of the communication channel and arrived at the useful concept of conditional entropy $H(X|Y)$. Its meaning is to enable one to estimate the average level of uncertainty of the random process X remaining if one has the possibility of observing the state of the random quantity Y. If X and Y are independent, then clearly an observation on Y says nothing about X and $H(X|Y) = H(X)$. On the other hand, if $X = Y$ (for example, when there is error-free transmission along the communication channel), then $H(X|Y) = 0$.

Thus, in the problem of transmission of information along a communication channel, the quantity $H(X|Y)$ can be treated as the average loss of information per transmitted value (per symbol or in unit time) in this communication channel. This means that it is natural to take $I(X;Y) = H(X) - H(X|Y)$ as the average measure of information that passes along the communication channel when the values of the random quantity X encoding the original messages are sent along it.The informative part of the messages is of no interest. We measure the information in bits and we measure the speed of its creation or reception in bits per symbol or in bits per unit time.

The values that a random quantity X can take can be considered to be the alphabet in which the messages subject to transmission are written (encoded). The messages are assumed to be long enough so that statistical characterizations can be used in general for the problem. This can be arranged alphabetically in different ways, as we have seen from the Shannon–Fano code. The optimal code for transmission is chosen with the characteristics of the communication channel to be used being taken into account.

The transmission capacity (3.7) of the communication channel is the maximal average speed of transmission along this communication channel that can be attained or that can be arbitrarily closely approximated in transmitting long texts of messages using sensible encodings beforehand in the channel alphabet.

3.4.5 *Calculating the transmission capacity of an abstract communication channel*

We have defined the capacity of an abstract communication channel by formula (3.7). We have just discussed the content of these ideas in broad outline. Now, in conclusion we nevertheless carry out a concrete calculation. We shall find the capacity of our abstract communication channel in the simplest example from which we began this abstract calculation. We recall the conditions.

In the communication channel the transmitter sends to the receiver the symbols 0 and 1. As a result of noise, the receiver occasionally deciphers the transmitted signal 0 as 1, and 1 as 0. Let p be the probability that the symbol is passed through correctly.

Along the channel messages (text, words) are sent consisting of sequences of letters, which are the symbols 0,1 of our very simple two-letter alphabet. We assume that the channel acts independently on each letter of the word, that is, it is a channel without memory.

What is the transmission capacity of such a communication channel?

In the present case the matrix of the probabilities of conversion is simplified to the limit not merely because we have a two-letter alphabet, but also because both transmitted symbols 0 and 1 have the same probability of being passed through without corruption. Thus the random quantity X at the input of the channel can take two values. Suppose that the encoding of the message to be sent is such that the probabilities of the appearance of the values 0, 1 are p_0, p_1, respectively.

At the output of the channel the random quantity Y can also take these two values 0 or 1, but possibly with different probabilities q_0, q_1. Let us find them.

The value 0 is obtained at the output with probability p when 0 is at the input, and with probability $1-p$ when 1 is at the input. In turn, at the input we have 0 with probability p_0 and 1 with probability p_1. Therefore the probability of getting 0 at the output is $pp_0 + (1-p)p_1$. Correspondingly, 1 appears at the output with probability $pp_1 + (1-p)p_0$.

The probability distribution of the combined random variable $Z = (X,Y)$ is also easy to write down: $(0,0) \sim pp_0$, $(0,1) \sim (1-p)p_1$, $(1,0) \sim (1-p)p_0$, $(1,1) \sim pp_1$.

We can now calculate the entropies $H(X)$, $H(Y)$, $H(X,Y)$ and via the second of formulae (3.9) find the speed of transmission of information. In our case we find that

$$I(X;Y) = H(Y) - h(p),$$

where $h(p) = -p\log p - (1-p)\log(1-p) = H(X,Y) - H(X) = H(Y|X)$.

The maximal value of the quantity $I(X;Y)$ is attained when $H(Y) = 1$, that is, when the distribution at the output is uniform: $q_0 = q_1 = \frac{1}{2}$. But $q_0 =$

$pp_0 + (1-p)p_1$, and $q_1 = pp_1 + (1-p)p_0$, therefore the condition $q_0 = q_1 = \frac{1}{2}$ holds precisely when the input distribution is uniform: $p_0 = p_1 = \frac{1}{2}$.

(Here we have used the following fact, which is easily verified: using the convexity of the logarithm function, if a discrete random quantity X has M different values, then $0 \le H(X) \le \log M$, where the left-hand relation holds with equality for the degenerate distributions when one value is taken with probability 1 and the others with probability 0, while the right-hand relation holds with equality for a uniform distribution.)

Thus, we have found that the channel transmission capacity of our simplest model communication channel with noise is equal to $C = 1 - h(p)$. Here $h(p) = H(Y|X)$ characterizes the loss of information on the symbol being transmitted. (See also [16].)

The calculation of the speed is, of course, always carried out to within a constant coefficient corresponding to the choice of the unit of time. For example (see [15]), suppose that the channel is physically capable of transmitting 100 bits 0, 1 in unit time, where each bit to be transmitted can be replaced by the opposite bit with probability 0,01. In this case, $h(p) = h(1-p) = h(0{,}01) \approx 0{,}0808$ and $C = 100(1 - 0{,}0808) = 91{,}92 \approx 92$ bits per unit of time. Take note the result is not equal to 99.

Armed with one's accumulated experience one can now try to prove the following intuitively clear Theorem of Shannon.

Theorem. *Suppose that there are a source of information X whose entropy per unit of time is equal to $H(X)$ and a communication channel of capacity C. If $H(X) > C$, then it is impossible to have an encoding delivering messages without delay or corruption. If, on the other hand, $H(X) < C$, then it is always possible to encode sufficiently long messages so that they are transmitted without delay; furthermore the probability of errors could be made arbitrariily close to zero.*

Part III
Classical Thermodynamics and Contact Geometry

Introduction

"This thermodynamics at present forms a remarkable scientific system the details of which, for all their beauty and shining completeness, do not yield to the entire system as a whole; it deserves the name of classical thermodynamics." This is what the great Lorentz wrote about classical thermodynamics when presenting his "Statistical theories in thermodynamics" [2].

Here we dwell on certain mathematical aspects of thermodynamics. We describe the two principles of thermodynamics in the language of differential forms (Chapter 1). We give an idea on the connection between classical thermodynamics and contact geometry (Chapter 2). Finally, we add an account of statistical physics and say a few words about the quantum-mechanical side of thermodynamics (Chapter 3).

Chapter 1
Classical thermodynamics (basic ideas)

1.1 The two principles of thermodynamics

1.1.1 Energy and the perpetual motion machine

Who does not know the words and phrases "energy", "law of conservation of energy" and "perpetual motion machine"? Many have even heard that a machine of this kind would be impossible.

But this is, in fact (after some refinement), the first of the two laws of thermodynamics, usually called the two principles of thermodynamics (and scientific philosophy).

The law of conservation of energy is part of our present consciousness. It is even difficult to believe that it appeared in such a form only in the second half of the 19th century (Mayer, Joule, Clausius, Helmholtz). For details see, for example, [4].

The most popular formulation of the first principle of thermodynamics consists exactly in the assertion that in nature there does not exist (and cannot be realized) a mechanism (source of mechanical energy, perpetual motion machine) whose single product of its cyclic activity would be carrying out work repeatedly and for ever (for example, lifting a load of 1 milligramme to a height of 1 millimetre).

1.1.2 Perpetual motion machine of the second kind and entropy

"It is well known that playing the basic role not only in this part of science but also in our general knowledge about the universe is the second principle of thermodynamics, or the Carnot–Clausius principle. In any case one can say that it reigns over more than half of physics." These also are the words of Lorentz in the book [2] already alluded to in the introduction.

The second principle of thermodynamics, although it is not a secret, is not so widely known to the man in the street as the first principle, although it has even more numerous formulations that are interesting in their dissimilarities of statement although they are equivalent in content. The simplest

V. Zorich, *Mathematical Analysis of Problems in the Natural Sciences*,
DOI 10.1007/978-3-642-14813-2_7, © Springer-Verlag Berlin Heidelberg 2011

and perhaps most banal (Clausius's formulation) states that when two bodies are in contact heat (energy) passes from the hotter body to the colder one (the first gets colder and the second gets hotter) and heat does not pass in the opposite direction by itself.

Another form of the second principle, in keeping with the formulation of the first principle given above, states that there does not exist a perpetual motion machine of the second kind, that is, a machine acting cyclically whose single result of a working cycle would be the conversion of heat energy taken from a heat reservoir (the ocean of heat) into mechanical energy (William Thomson, who in 1892 became Lord Kelvin in recognition of his services to science).

Without going deeply here into the details of thermodynamics, which are well known and presented in any respectable textbook, we merely add that, strangely enough, historically the second principle was formally discovered before the first one by the father of thermodynamics Sadi Carnot (1824). He answered the question posed by James Watt (1765) on the possible efficiency of a heat (steam) engine. (This was the period of the appearance and dissemination of steam engines and motors. What Watt asked was quite specific: how much coal is required for such a machine to carry out a given amount of mechanical work.)

The work of Carnot (forgotten and discovered in 1834 by Clapeyron) was developed in the 1850s by Clausius; he brought into existence the second fundamental concept of thermodynamics (the first being energy), namely the *entropy* of a thermodynamic system, introduced by him in 1865.

In a certain sense the two principles of thermodynamics produced as a result the two fundamental characteristics of the state of a thermodynamic system: its *energy* and its *entropy*.

Any closed (isolated from external influences) thermodynamic system evolves in the direction of increase of the entropy of its state. For example, a gas released from the vessel containing it diffuses throughout the whole room, that is, it goes from a more organized state (collected in the vessel) to a less organized state (more likely than if it had suddenly again collected in the vessel)[1].

At the end of the 19th and beginning of the 20th centuries, on the one hand statistical thermodynamics (Maxwell, Boltzmann, Gibbs, Poincaré, Einstein) was born; on the other hand there began the mathematical description and formalization of classical thermodynamics (Maxwell, Gibbs, Planck, Carathéodory).

---

[1] Sommerfeld [6], stressing the role of the second principle, quotes Robert Emden "In the gigantic fabric of natural processes the principle of entropy takes the role of the director who gives instructions about the form and flow of all transactions. The law of conservation of energy merely plays the role of the bookkeeper who keeps debit and credit in balance".

1.2 The two principles of theormodynamics in a mathematical setting

Every science has its own favourite toy. In thermodynamics it is a gas in a cylinder under a piston. The piston can move changing the volume of the gas. The walls of the cylinder can conduct heat and, conversely, they can isolate the gas from heat exchange with the external medium.

Playing on paper in this apparatus (key elements of both the steam engine and the modern internal combustion engine) Carnot carried out one of the first ingenious (and not costly) thought experiments in physics. By moving the piston and heating or, when required, cooling or thermally isolating the cylinder he devised a cyclic process now called (after some modifications made by Clausius) the Carnot cycle. Carnot found the answer to Watt's question on the possible coefficient of useful action of any heat engine and at the same time, as it turned out, made the great discovery, which (after development by Clausius) became the second principle of thermodynamics. (The idea of the first principle is essentially also present in Carnot's arguments.)

We too can play a little. This will enable us both to accept useful mathematical schemes and also not to lose the physical content in the abstractions which arise.

The classical Clapeyron's law states that the volume V of a gas, its pressure P and its temperature T are connected by the formula $\frac{PV}{T} = C$ (called the equation of state) characterizing the equilibrium thermodynamic state of the gas. Here the constant C depends only on quantity of gas; see [7], [8a].

If we push on the piston little by little (so as not to depart from the equilibrium state), by changing the volume of the gas by some quantity dV (which is negative if we compress the gas), then we perform work $-PdV$. Part of it goes into increasing the energy E of the gas (which we have compressed like a spring) and the part in the form of heat δQ can escape into the external medium if our cylinder conducts heat. If, on the other hand, it is adiabatic, that is, no heat is conducted, there is no heat exchange with the external medium, no heat is lost ($\delta Q = 0$) and the internal energy E of the gas is simply increased precisely by the amount $dE = -PdV$ of work done by us. One naturally assumes that the gas itself can also perform work $\delta W = PdV$. (For example, if it had pushed the piston increasing its volume by $dV > 0$, then clearly it would perform work $\delta W = PdV$.)

In the general case the energy balance is such that

$$\delta Q = dE + \delta W. \tag{1.1}$$

We observe that, by contrast with dE, the differential forms δQ and δW are not exact. They are not differentials of functions. The work required, for example, to double the volume of a gas depends not only on the initial and final values of the volume, but also on the integral heat exchange with external medium, which arises here. Both these quantities essentially depend on the conditions under which the transition from one thermodynamic state to the other is carried out. For example, if the process is adiabatic, then there is no heat exchange at all. In such a process (transition via this path) the integral of the form δQ is equal to zero. Via the other path joining the same thermodynamic states the integral of the form δQ will generally be non-zero if the walls of the cylinder conduct heat. Clearly the same automatically relates to the differential form of work δW. For this reason we are required to use different symbols for the differential in the fundamental equation 1.1.

Everything that has been said so far (without discussion or precise formulations) has had the aim of reviving in the memory of the reader the most general information in thermodynamics. It serves as a motivation for the formal definition of a thermodynamic system which we give below.

1.2.1 Differential form of heat exchange

A landmark advance in science is often achieved (or, to put it better, taken shape) by an interesting and characteristic method particularly showing up in abstract areas such as theoretical physics and mathematics.

Imagine an hour glass. In order for it to work it has to be turned "from foot to head" from time to time. In mathematics it is the same. First one obtains several new interesting facts. Among them is revealed something that in some or other respect is central and key, connecting a lot of earlier material. This is taken as an original principle turning everything from foot to head (for example, by making a theorem an axiom) and development continues by relying now on this new principle, heralding a large area of facts of mathematics and the universe.

For example, Newton's laws were not developed in an empty place (we recall Kepler, Galileo or the burnt-down library of Alexandria, whose scrolls contained (according to some sources) the heliocentric system and Kepler's laws and many of the foundations of modern mathematics). Newton's laws generalized an enormous amount of experimental material. By taking Newton's laws as the foundations we can obtain from them much of what gave birth to them. The subsequent development of physics led to new variational principles of mechanics, including a large area of phenomena and interactions different from the interactions described by central forces.

At such turning points there occurs, if one can express it in this way, a change of scale. Here it consists exactly in a change and lessening of the number of principles in the simultaneous extension of the field of objects and phenomena that they envelop and unite.

At the end of the 19th century and beginning of the 20th century, via the work of Maxwell, Gibbs, Planck, Carathéodory and other scholars the rich material of thermodynamics underwent a mathematical systematization. As usually happens in a sufficiently advanced formalized theory, there arose the fundamental statements, concepts, principles, axioms and theorems concisely containing the numerous concrete achievements obtained by the work of many researchers. And it is here that we turn to such a moment of turning the hour glass upside down.

We suppose that the equilibrium state of our thermodynamic system is determined by a set of parameters $(\tau,a_1,\dots a_n) =: (\tau,a)$, where $\tau > 0$ will play the role of *absolute temperature* and $a = (a_1,\dots a_n)$ is a set of *external parameters* which can be changed like the volume of a gas under a piston in the "toy" example considered above.

In the model that we consider the thermodynamic system itself is identified with the fundamental differential form

$$\omega := dE + \sum_{i=1}^{n} A_i da_i, \tag{1.2}$$

called the *heat-exchange form* or the *heat-influx form*. Here, by definition, E is the *internal energy of the system* and A_i is the *generalized force* corresponding to the variation of the coordinate a_i (that is, $\sum_{i=1}^{n} A_i da_i$ corresponds to the work δW of the system associated with the change of the external parameters and the form ω itself corresponds to the heat-exchange differential form δQ of equation (1.1)). The quantities E and A_i naturally depend on $(\tau,a_1,\dots a_n)$. These dependences enter into the definition of a thermodynamic system. Formally, they actually constitute this definition. The relations $A_i = f_i(\tau,a_1,\dots a_n)$ are called the *equations of state.*

The form ω defining the thermodynamic system must satisfy certain requirements, which we now point out.

1.2.2 The two principles of thermodynamics in the language of differential forms

Consider an oriented path γ in the space of equilibrium states (τ,a) corresponding to the transition of the sytem from one equilibrium state (τ_0,a_0) to another (τ_1,a_1). Then the integrals

$$\int_\gamma \omega = \int_\gamma \delta Q, \qquad \int_\gamma \sum_{i=1}^{n} A_i da_i = \int_\gamma \delta W$$

give, respectively, the amount of heat received by the system and the work of the system associated with the external forces in such a transition, and the integral of dE (see (1.1), (1.2)) gives the change in internal energy of the system.

The second principle of thermodynamics, discovered by Carnot, by the work of Clausius reduces to the assertion that for any realizable thermodynamic cycle γ the remarkable inequality

$$\int_\gamma \frac{\delta Q}{T} \geq 0$$

holds, where T is the real physical absolute temperature and the equality

$$\int_\gamma \frac{\delta Q}{T} = 0$$

is realized if and only if the closed path γ is in the space of equilibrium states of the thermodynamic system.

This means that the differential form $\frac{\delta Q}{T}$ is the differential of some function S of the equilibrium state of the system.This is what Clausius called the *entropy* of the thermodynamic state of the system. For now the function is defined to within an additive constant. Most often one merely needs the difference of the values of this function in going from one state to another. In this case the additive constant plays no role. (There are, however, reasons for supposing that at $T = 0$ the entropy of the system is the same for all states. This is the Nernst heat theorem or, as they say, the third principle of thermodynamics. Therefore at $T = 0$ one sets $S = 0$. Here we shall not dwell on this and various other matters. See page 108.) Thus, the absolute temperature T as a function of the state of the thermodynamic system is remarkable in that T^{-1} is the integrating factor for the differential form δQ of heat exchange, converting it to the exact form, namely, the differential of the function S, which is the entropy of the system.

Formalizing what we have said above, we assume that in the mathematical model defining the thermodynamic system the form ω is such that *the form $\tau^{-1}\omega$ is exact*. Naturally, we call the function S for which $dS = \tau^{-1}\omega$ the *entropy* of the system.

We now consider an adiabatic process. Now it is a pathγ in the space of equilibrium states (τ, a) going along the null spaces $(\ker \omega)$ of the differential form ω. (In the old notation this reduces to the fact that $\delta Q = 0$ along γ, that is, there is no heat exchange with the external medium.) Then

$$\int_\gamma \omega = \int_\gamma dE + \int_\gamma \sum_{i=1}^{n} A_i da_i = E_1 - E_0 + \int_\gamma \sum_{i=1}^{n} A_i da_i = 0, \tag{1.3}$$

which is the law of conservation of mechanical energy, which means the first principle of thermodynamics.

Note that formula (1.3) holds in general for any process in which the total heat exchange with the external medium, expressed by the integral on the left-hand side is equal to zero. For example, this is the case if γ is a closed curve (a cycle) in which the temperatures of the thermodynamic system and the external medium are constant and the same.

Completing the definition of the mathematical model of a thermodynamic system as the differential form (1.2), we now transform the first principle as the requirement that *the form of the work* $\sum_{i=1}^{n} A_i da_i$ *for constant values of the temperature* τ *be closed.* (In view of Poincaré's lemma, it is then exact in any simply connected region of the parameters a. Hence a closed work cycle of any mechanical machine working in such a region of the parameters at one and the same temperature cannot be a source of mechanical energy — a perpetual-motion machine.) Thus the simplest mathematical model of a thermodynamic system has been indicated.

Differentiating the form (1.2) and taking the above into account we get

$$d\omega = d\sum_{i=1}^{n} A_i da_i = \sum_{i=1}^{n} \frac{\partial A_i}{\partial \tau} d\tau \wedge da_i \,. \tag{1.4}$$

If we recall that $\omega = \tau dS$, then from formula (1.4) with $1 \leq i \leq n$ we obtain

$$\frac{\partial A_i}{\partial \tau} = \frac{\partial S}{\partial a_i} \,. \tag{1.5}$$

1.2.3 *Thermodynamics without heat*

From the formal-logic point of view, as we learnt in school, no special "caloric" weightless fluid flowing from body to body exists, that heat is energy, that there is a mechanical equivalent of heat enabling one to forget about calories, and so on, it is natural after this to construct a formal theory without the redundant dependent concepts such as the heat featuring on the left-hand side of the fundamental equation (1.1).

Namely, the right-hand side of formula (1.1) and only it defined the differential form in formula (1.2), which was the starting point of the mathematical constructions. In formula (1.2), by contrast with formula (1.1), we do not have an equality, but the definition of the form ω. This form is expressed in terms of the energy of the system and its work, that is, only in terms of energy.

The integral of this form along any path is now already by definition called *the flow of heat* or *the heat* obtained by the system from the external media under the indicated change of state of the thermodynamic system.

1.2.4 Adiabatic transition and Carathéodory's axiom

Adiabatic processes, which occur without heat exchange with the external medium, have a special role in these constructions. This, as we have already remarked in the derivation of formula (1.3), is equivalent to the property that the path γ in the state space proceeds tangent to the kernels $\ker \omega$ of the differential form ω at the moving point. In other words, the velocity vector of the motion is always in the corresponding space of $\ker \omega$.

Recalling that $\omega = \tau dS$ we can conclude that adiabatic processes take place without the value of the entropy being changed. The path γ lies on a level surface of the function S. Furthermore, the distribution itself of the kernel $\ker \omega$ is tangent to the entropy-level surface.

Thus it is impossible to pass adiabatically through equilibrium states from one thermodynamic state of the system to another if these states have different entropy levels.

In his axioms of classical thermodynamics Carathéodory [11a] took this circumstance as the initial equivalent of the second principle of thermodynamics. Namely, by introducing the form (1.2) instead of the mathematical requirements given above he proposed a thesis more familiar to physicists: *in any neighbourhood of an equilibrium thermodynamic state of a system there are equilibrium states into which it is impossible to pass by an adiabatic process.*[2]

After this the mathematics starts, which leads to what was set forth above. This mathematics has a geometric flavour and is interesting in its own right because of the problems it poses and solves. This will be talked about in the next chapter.

But here we merely add the following, which is known to physicists but may be unknown to some mathematicians (not Carathéodory). Slow equilibrium adiabatic transitions between thermodynamic states in fact preserve entropy and are reversible. But non-equilibrium transitions between equilibrium states can also be completed without heat exchange. For example, suppose that a gas is in half a thermally isolated vessel (a thermos) partitioned by a barrier. You now remove the barrier and the gas fills the whole vessel. After a while it settles down to its new equilibrium state. It can be shown that the entropy of this state is larger than the entropy of the original equilibrium state of the gas when it occupied only half the volume of the vessel.

We shall calculate the entropy of an ideal monatomic gas. We recall the Clapeyron (Clapeyron–Mendeleev) law, which is well-known from school:

$$P = \frac{n}{N}\frac{RT}{V} = nk\frac{T}{V},$$

[2] A violation of this proposition (axiom) leads almost immediately to the construction of a perpetual-motion machine of the second kind.

where P is the pressure, V the volume, T the absolute temperature of the gas; R is the universal gas constant, N is the Avogadro number, n is the number of molecules of the gas and $k = \frac{R}{N}$ is the Boltzmann constant.

We also recall that the average kinetic energy of a molecule of a gas is equal to $\frac{3}{2}kT$, therefore the internal energy E of the whole gas is equal to $\frac{3}{2}nkT$ and $T = (\frac{3}{2}nk)^{-1}E$. Now, taking into account the fact that $\delta Q = dE + PdV$ and $dS = \frac{\delta Q}{T}$, we find that

$$dS = (\frac{3}{2}nk)\frac{dE}{E} + nk\frac{dV}{V},$$

and hence, to within an additive constant, we obtain

$$S = nk\ln(VE^{3/2}) = k\ln(V^n E^{\frac{3}{2}n}) .$$

In particular, if while preserving the internal energy of the gas we allow it to occupy twice its volume, then the entropy of the new state will increase by $nk\ln 2$.

It is interesting that it is no longer possible to return to the earlier state without heat exchange, even by violating the equilibrium of the intermediate states.

Thus Carathéodory's axiom could also apply to non-equilibrium transitions carried out without heat exchange. Incidentally, without them it is apparently even impossible to give an adequate sufficiently complete decription of equilibrium thermodynamics (see, for example, [11b]).

Chapter 2
Thermodynamics and contact geometry

2.1 Contact distributions

2.1.1 *Adiabatic process and contact distribution*

Carathéodory's axiom given above states the physical law touching on adiabatic transitions between thermodynamic states. We shall give a full mathematical formulation of it and analyse the problems and after-effects it leads to.

We have the differential 1-form ω defined by formula (1.2) of Chapter 1.

In each tangent space to the space of states of the thermodynamic system there is the hyperplane $\ker\omega$, at the vectors of which the form ω vanishes. These tangent hyperplanes $\ker\omega$, which are the kernels of the form ω, are distributed throughout the whole space M of states. We denote this *distribution* by the symbol H. This gives rise to the pair (M, H) of the space and the distribution of tangent planes. A path γ in M is considered to be *admissible* if at each of its points it is tangent to the corresponding plane of the distribution H.

Question: Is it always possible to join two points of M by an admissible path? (We assume that M is a smooth connected manifold.)

Carathéodory's axiom asserts that the corresponding thermodynamic form ω is such that in a neighbourhood of any point of the space M there are points that cannot be accessed by an admissible path from our point.

2.1.2 *Formalization*

Forgetting thermodynamics for the moment, we now turn to purely mathematical concepts, structures, objects and questions arising in this connection.

Let M be a smooth connected n-dimensional manifold and let H be the distribution of tangent hyperplanes (of dimension $n-1$) on it. Planes of the distribution H are often called *horizontal* and *admissible* paths going along them are often also called *horizontal* or *integral*.

For example, if some group of diffeomorphisms acts transitively and freely (that is, without fixed points) on M, then the tangent hyperplane at

DOI 10.1007/978-3-642-14813-2_8,

a point spread along M by the transformations of the group generate a distribution H in M. A differential 1-form ω on M that is non-vanishing everywhere generates the distribution $H = \{\ker\omega\}$ on M, as we have seen.

2.2 Integrability of distributions

2.2.1 *The Frobenius theorem*

A distribution H on M is said to be *integrable* if it has an integral surface, that is, a submanifold such that at each of its points the corresponding plane of the distribution H is a tangent plane of the submanifold.

For example, if the distribution H were 1-dimensional, more precisely, defined by a vector field, then by the fundamental theorem on the existence of solutions of differential equations it would always be integrable (an integral curve passes through each point).

In the general case this is not so, as is well known. An example is: $H = \{\ker\omega\}$, where $\omega = y dx + dz$ in $\mathbb{R}^3$ (corresponding to a circuit of a closed contour in the (x,y)-plane is a non-closed integral curve in $\mathbb{R}^3$ — the "coil of a spiral" with pitch equal to the area enclosed by the contour). The distribution of tangent hyperplanes $H = \{\ker\omega\}$ generated by a non-degenerate smooth differential 1-form ω is integrable only when $d\omega = 0$ on the corresponding planes $\ker\omega$.

This condition is equivalent to the following condition of Frobenius for the integrability of the distribution $H = \{\ker\omega\}$ in $\mathbb{R}^3$:

$$\omega \wedge d\omega = 0.$$

For a distribution of k-dimensional planes given on an n-dimensional manifold in the form of the common zeros of $n-k$ smooth differential forms $\omega_1,\ldots,\omega_{n-k}$ that are everywhere independent, that is, when the planes of this distribution H are obtained as the intersection of the hyperplanes $\ker\omega_1,\ldots,\ker\omega_{n-k}$, the criterion for the integrability of the distribution H consists in the requirement that $d\omega_i = 0$ for the planes of the distribution H for each of the forms ω_i. This is equivalent to fulfilment of the following conditions (see, for example, [4]):

$$\omega_1 \wedge \ldots \wedge \omega_{n-k} \wedge d\omega_i = 0\,, \qquad 1 \leq i \leq n-k\,.$$

2.2.2 *Integrability, connectibility, controllability*

We now turn to the question whether points of a space can be joined by an admissible path. This question arises in control theory where the word "connectibility" is replaced by the term "controllability".

For us the question arose in connection with thermodynamics, adiabatic processes and Carathéodory's axiom. This question relates directly to the integrability of the hyperplanes of the corresponding distribution H. Namely, integrability and connectibility are complementary to each other: integrability is non-connectibility and connectibility is non-integrability.

In fact, if the distribution H of the hyperplanes is integrable, then any admissible path tangent to it cannot leave an integral surface of the distribution. Hence in a neighbourhood of any point there are points of the manifold that are inaccessible from it by an admissible path.

Carathéodory proved that for a distribution of hyperplanes the converse statement holds locally: non-connectibility implies integrability [11a].

(Then he proves that a form ω defining the distribution $\ker \omega$ locally admits an integrating factor, that $\tau^{-1}\omega = dS$, that S is the entropy and the second principle of thermodynamics holds.)

The mathematical content of the work of Carathéodory apparently stimulated geometers interested in physics, such as P.K. Rashevskiĭ, to investigate this useful geometric question of the connectibility of points of a space by paths that are admissible for the given distribution.

The first general answer in terms of Lie brackets was obtained by Rashevskiĭ in the paper [5], but is called Chow's theorem; see [6], [7]. People in the know refer to it as the "Rashevskiĭ–Chow theorem". The study of this question, its extension and its connection with aspects of control theory continue to this day; see, for example, [7], [8].

In terms of Lie brackets the conditions of integrability and connectibility are formulated in the following way (see, for example, [9], [10], [11]). Let $e_1, \ldots, e_k$ be smooth vector fields on a manifold M that are linearly independent at each point and generate a distribution H of the family of k-dimensional tangent planes. The distribution H is integrable if and only if the Lie brackets $[e_i, e_j]$ of the fields generating the distribution do not go out beyond the plane of the distribution.

In order that locally (and, on a connected manifold, globally) any points of the manifold be connectible by a path that is admissible for the distribution H it suffices that the iterated brackets $[[e_i, e_j] \ldots]$ of the original field $e_1, \ldots, e_k$ generate a basis of the entire tangent space to the manifold at each point of the manifold.

These formulations and Frobenius's conditions are connected by the following well known formula in the calculus of differential forms. (In general, the language of differential forms is dual to the language of vector fields.) If X and Y are smooth vector fields and ω is a 1-form on the manifold, then

$$d\omega(X,Y) = X\omega(Y) - Y\omega(X) - \omega([X,Y]),$$

where $X\omega(Y)$ and $Y\omega(X)$ are the Lie derivatives of the functions $\omega(Y)$ and $\omega(X)$ along the fields X and Y, respectively, and $[X,Y]$ is the bracket (commutator) of these vector fields.

In the general case of a form ω of order m we have

$$d\omega(\xi_1,\ldots,\xi_{m+1}) = \sum_{i=1}^{m+1}(-1)^{i+1}\xi_i\omega(\xi_1,\ldots,\widehat{\xi_i},\ldots,\xi_{m+1}) + \\ + \sum_{1\leq i<j\leq m+1}(-1)^{i+j}\omega([\xi_i,\xi_j],\xi_1,\ldots,\widehat{\xi_i},\ldots,\widehat{\xi_j},\ldots,\xi_{m+1}),$$

where $\widehat{\ }$ denotes a missing term, $[\xi_i,\xi_j]$ are the Lie brackets of the fields ξ_i, ξ_j and $\xi_i\omega$ is the derivative of the function $\omega(\xi_1,\ldots,\widehat{\xi_j},\ldots,\xi_{m+1})$ with respect to the field ξ_i.

2.2.3 *Carnot–Carathéodory metric*

Let M be a Riemannian manifold with metric g. Suppose that the distribution H in M is such that any pair of points in M can be joined by a curve that is admissible for H. Then one can introduce in M (more precisely, in (M,g,H)) a new metric in which the distance between points is measured by the infimum of the lengths of admissible curves joining these points.

This metric (apparently from the light hand of M. Gromov) is called the *Carnot–Carathéodory metric* or the CC-*metric*.

It arises and is useful in various situations where triples of the form (M,g,H) appear. For example, in a complex space complex tangents to non-trivial hypersurfaces (for example, to a hypersphere) form non-integrable contact distributions.

It is helpful to recall that in view of Darboux's theorem any *contact form* (that is, a smooth form ω such that the form $d\omega$ is non-degenerate on $\ker\omega$) can, by a smooth change of coordinates, be reduced to the form $x_1dy_1 + \ldots + x_ndy_n + dz$. For example, in the case $\mathbb{R}^3$ we obtain the form $xdy + dz$.

Note that here motions along lines parallel to the x axis are admissible. In the planes $x = c$ motions along sloping lines are admissible. The slope is linearly dependent on c. After that observation it becomes obvious that it is possible to join any two points of the space $\mathbb{R}^3$ by a path that is admissible for the distribution $H = \{\ker(xdy + dz)\}$.

Of course, the non-integrability of a distribution formed by the contact form $xdy + dz$ can be verified by referring to Frobenius's condition. Concerning contact structures see, for example, [9], [10], [12], [13].

2.2.4 The Gibbs contact form

We already know the relation $\omega = \tau dS$ for the form (1.2) of Chapter 1, which is fundamental and corresponds to the second principle of thermodynamics. Here ω has the meaning of the form of heat exchange, τ is the absolute temperature and S is the entropy.

After Gibbs we consider the form

$$\Omega = -\tau dS + \sum_{i=1}^{n} A_i da_i + dE,$$

or, more specifically (as Gibbs wrote it), the contact form

$$\Omega = -TdS + PdV + dE,$$

with the customary terminology for classical thermodynamics: T the absolute temperature, S the entropy, P the pressure, V the volume and E the internal energy of the gas, liquid or other thermodynamic system.

The fundamental relation (1.1) of Chapter 1, which expresses the energy balance, where now $\delta Q = TdS$, states that equilibrium thermodynamic processes (no longer necessarily adiabatic) take place in such a way that $\Omega = 0$ on them, that is, they go along the distribution $\{\ker \Omega\}$ in the space $\mathbb{R}^5$ of parameters (T,S,P,V,E). But this is not the space M of states (τ,a), or (T,V), or, for example, (S,V).

In fact, only two parameters remain free and in reality equilibrium processes flow on a *Legendre surface* (integral submanifold of M of maximal dimension) *of the contact structure* defined by the form Ω in the space $\mathbb{R}^5$. This is one of Gibbs's main theses.

In the case of the general form

$$\Omega = -\tau dS + \sum_{i=1}^{n} A_i da_i + dE \,,$$

we have $2n+3$ variable quantities $(E,S,A_1,...,A_n,\tau,a_1,...,a_n)$, whereas the state of the system is determined only by the parameters $(\tau,a_1,...,a_n)$ (or by a set of $n+1$ independent quantities equivalent to them). This means that in the general case equilibrium thermodynamic processes go along Legendre $(n+1)$-dimensional manifolds (integral manifolds of maximal dimension) of the distribution of hypersurfaces $H = \{\ker \Omega\}$, which in $\mathbb{R}^{2n+3}$ is generated by the (contact) form Ω.

2.2.5 *Concluding remarks*

In concluding this section we observe that here we have only been concerned with one specific geometric question relating to classical thermodynamics, although it is true that we have also talked about how, in modern mathematical language, the fundamental principles of thermodynamics are formulated.

There are also other directions of modern mathematical investigations generated by classical thermodynamics, for example, the attempt to formalize (axiomatize) the notion of entropy [14] and the generalization of the procedures connecting the basic thermodynamic functions (potentials) [15].

However, the fundamental physically interesting mathematical layer opened up by Gibbs [1a], [1b], [1c] relates to contact geometry. An idea of this can be obtained by looking at the volume of works at the conference devoted to the 150th birthday of Gibbs [2]. It took place at the same Yale University where Gibbs himself worked.

I cite just the following sentence of the article [3] by V.I. Arnold placed in this collection: " I think that Gibbs was the first person who understood the significance of contact geometry for physics and thermodynamics".

The fundamental thermodynamic equality

$$dE = TdS - PdV$$

in 1873, when Gibbs's work appeared, had undoubtedly already become the standard basis in the construction of thermodynamics. But even Clausius, who in 1865 formulated the second principle of thermodynamics, conceptually defined entropy in 1854 and gave it its name in 1865, never gave it a central role in the exposition of thermodynamics. For Clausius and his contemporaries thermodynamics reduced to the interconnection between heat and work.

Gibbs was the first to exclude heat and work from the foundations of thermodynamics by giving preference to the functions of thermodynamic state: energy and entropy.

Thermodynamics became the theory of properties of matter in a state of thermodynamic equilibrium. In particular, Gibbs showed that for the equilibrium of two phases 1 and 2 of one and the same substance it is not enough to have equal temperatures and pressures. In addition it is necessary that the energies, entropies and volumes of the phases satisfy (relative to the unit of mass) the relation: $E_2 - E_1 = T(S_2 - S_1) - P(V_2 - V_1)$, and an equilibrium state of an isolated thermodynamic system is stable if and only if it satisfies a local maximum of the entropy.

Gibbs strove to find a general graphical (geometrical) method that would in the large describe thermodynamic properties of a medium (in all its phases) for reversible processes.

In 1873 Gibbs introduced his famous fundamental surface connecting the energy E of a thermodynamic system with its entropy S and volume V: $E = E(S, V)$. He showed how the nature of its tangent planes was related to the coexistence of the phases and critical points.

This work was admired by Maxwell and stimulated him to construct his well-known Maxwell model of this surface. Gibbs showed that $E(S, V)$ is the generating (characterisic) function whose knowledge determines all the thermodynamic properties of the system.

They noted that, just as Lagrange's "Analytical mechanics" was the crown of eighteenth-century science, Gibbs's memoir of 1876,1878 "On the equilibria of heterogeneous media" [1c] was to a significant extent the crown of classical natural science in the nineteenth century.

Chapter 3
Thermodynamics classical and statistical

3.1 Kinetic theories

The original idea of statistical physics was to obtain thermodynamics from mechanics.

"The honour goes to the Austrian physicist Boltzmann for being the first to successfully approach this problem and establish the connection between the notion of probability (defined in a reasonable fashion) and the thermodynamic functions, in particular, entropy. Along with him one must also consider Willard Gibbs to be one of the founders of this new branch of theoretical physics — statistical thermodynamics. Further one must recall the work of Poincaré, Planck and Einstein". (Lorentz 1915, [5])

3.1.1 Molecules and pressure

The origin of the kinetic theory of gases is due to Daniel Bernoulli. In his book "Hydrodynamics" (Strasburg, 1978) the pressure of a gas had already been derived from the change of momentum of the molecules of a gas colliding with the walls of the vessel. In 1746 Daniel and Johann Bernoulli were even awarded a prize of the Paris Academy of Sciences for work on this topic.

Further progress in kinetic theory and statistical thermodynamics relates to the second half of the 19th and the beginning of the 20th centuries and is associated with the names of Clausius, Maxwell, Boltzmann, Gibbs, Poincaré and Einstein.

3.1.2 The Maxwell distribution

One could naïvely assume that the molecules of a gas in a state of thermodynamic equilibrium all enjoy approximately the same speed (kinetic energy) as a result of numerous collisions and redistribution of the momenta. Even more impressive was the discovery of Maxwell (1866), who showed

V. Zorich, *Mathematical Analysis of Problems in the Natural Sciences*,
DOI 10.1007/978-3-642-14813-2_9,

by simple arguments that the projections of the velocities of the molecules of an ideal gas onto any direction are normally distributed. Accordingly this gives rise to the classical Maxwell's distribution of molecules according to their kinetic energies.

3.1.3 *Entropy according to Boltzmann*

Boltzmann (1868–1871) developed Maxwell's result and showed that the equilibrium energy distribution of the particles of an ideal gas in an external field of force (for example, in a field of gravity) is determined by a distribution function $\exp(-E/kT)$, where E is the sum of the kinetic and potential energies of a particle, T is the absolute temperature and k is the Boltzmann constant.

The relation $S = k \log \Pi$ (discovered by Boltzmann in 1872) between the entropy S and the probability (understood in a certain sense) of the thermodynamic state[1] is called Boltzmann's principle or Boltzmann's formula. It gives a statistical interpretation of the second principle of thermodynamics, which in the end reduces to the assertion that a thermodynamic process has a tendency to transfer a system from a less probable state to a more probable one, that is, in the direction of increasing entropy. A conditional maximum of the entropy function S corresponds to an equilibrium state.The main achievement of Boltzmann consists in the successive realization of the idea of interpreting classical thermodynamics as statistical mechanics; see [1].

He derived the fundamental kinetic equation for a gas, determining (under certain assumptions) its evolution as a mechanical system and obtained his remarkable H-theorem stating that the evolution proceeds in the direction of increasing entropy.

Boltzmann also pointed out the (now widely known) expression

$$H = -\int \varrho \log \varrho$$

for the entropy in terms of distribution density, which, along with the notation H (from the Greek word $\eta\nu\tau\rho o\pi\iota\alpha$ and the Greek letters η, H) migrated to other spheres, becoming, for example, a fundamental notion of the modern theory of transmission of information. (By contrast with Clausius, Boltzmann apparently first denoted entropy by the letter E, which has now been given back to energy. We now accept that entropy is denoted by S, leaving the symbol H for the Hamiltonian of a Hamiltonian mechanical system.

[1] Here Π is the number of microstates of the system, for example, the states of molecules of a gas in a given container corresponding to the macrostate whose entropy is S. The modern value of Boltzmann's constant $k = (1{,}38044 \pm 0{,}00007) \cdot 10^{-23}\mathrm{J/K}$.

3.1.4 *The Gibbs ensemble and the "thermodynamicization" of mechanics*

Gibbs proposed the most general and orderly mathematical basis for statistical mechanics and thermodynamics.

Sommerfeld, who was not prone to becoming enraptured, called Gibbs a "great thermodynamicist and statistician". His first works, after being hidden away in the Works of the Academy of the State of Connecticut, became widely known after Ostwald in 1902 published them in German under the title "Studies in thermodynamics" ([17], p. 67).

In the year up to the time of his death in 1902 Gibbs published his "Statistical mechanics", which served as the mathematical foundation for subsequent generations; see [2].

As we have already observed, Gibbs combined the ideas of thermodynamics and Hamiltonian mechanics. He introduced the remarkable mathematical structure of statistical physics, namely, a Hamiltonian system endowed with a probability measure evolving under the action of the Hamiltonian flow in the phase space of the system. This model became the source of numerous problems and investigations of the theory of dynamical systems, which is actively carried out even today.

We briefly recall that in the phase space Γ of a Hamiltonian system Gibbs introduced what is now called the *canonical* distribution of probability states. The density ϱ of this distribution is defined by the Hamiltonian (energy) $H = H(q,p)$ of the system

$$\varrho := c \exp(-\beta H),$$

where $c = (\int_\Gamma \exp(-\beta H) dq dp)^{-1}$ is a normalizing factor and in physical systems $\beta = 1/k\tau$, k is the Boltzmann constant and τ is the absolute temperature.

(A significant part of the book [12] is devoted to a detailed discussion and substantiation of the canonical distribution.) The canonical distribution is invariant with respect to the action of the Hamiltonian flow, since the measure and the Hamiltonian are invariant.

The Hamiltonian can depend on parameters $a := (a_1, \ldots, a_n)$, that is, $H = H(q,p,a)$. These can be parameters of external-action type such as shifting a piston or changing the volume of a gas.

In particular, Gibbs pointed out the following beautiful natural process of "thermodynamicizing" a Hamiltonian mechanical system. Consider the average energy

$$E(\beta,a) = \int_\Gamma H \varrho dq dp$$

and the averaged forces of the bonds' reactions corresponding to the parameters a_i,

$$A_i = \int_\Gamma \frac{\partial H}{\partial a_i} \varrho dq dp .$$

The last relations will be interpreted as the equations of state $A_i = f_i(\beta, a)$, $i = 1, \ldots, n$. It is not difficult to verify that the 1-form

$$\omega = dE + \sum_{i=1}^{n} A_i da_i$$

satisfies the axioms of the two principles of thermodynamics.

3.1.5 Ergodicity

The original idea of statistical physics, as has already been recalled above, was to obtain thermodynamics from mechanics. For example, by regarding a gas in a vessel as a mechanical system consisting of an enormous number of weakly interacting chaotically moving particles, one can find its internal energy, associate the average kinetic energy of the particles with the temperature of the gas and find the pressure and the entropy.

Here the functions of classical thermodynamics are obtained as the statistically average quantities arising as a result of summing the enormous number of small contributions of the individual particles of this multiparameter mechanical system. It is here that probabilistic ideas, the laws of large numbers and the concentration principle start to work. This is a new ideology. In it the second law of thermodynamics (and much else) is no longer regarded as an absolute truth, by virtue of which the system by itself never goes over into a state of smaller entropy, or heat by itself never goes from a cold body to a hot one. Now all the molecules can, in principle, collect in a corner of a room; but the probability of this event proves to be negligible, as also is the time of its possible existence. In practice we are not in a position to observe this event.

There arise many fundamental questions. For example, what, properly, is it that we observe and how do we measure it?

When one wants to think about all possible states of a given system, one can either trace its evolution in time, or one can imagine an entire *ensemble* of systems which are copies of our system taken at one and the same moment of time and representing all possible states of our system.

From the point of view of mechanics such an ensemble is simply the phase space of the system in which each individual point represents the the system in some specific state. The evolution of the system is depicted by the trajectory of a point in the phase space. The motion of points induces a motion of the whole phase space, aptly called the *phase flow*.

Liouville's classical theorem states that *the phase flow of a Hamiltonian system preserves the phase volume*. That is, it flows like an incompressible fluid.

It is therefore natural to assume that the probability of finding the system in one of the states associated with some region of the phase space accessible to the motion is proportional to the volume of this region.

On the other hand, one can trace the whole evolution of an individual system and the probability of a state is assumed to be proportional to the time during which it exists.

In this connection Boltzmann even initially assumed that the system in its evolution runs through all possible states on the level surface of given energy and therefore that both approaches to the measurements of averages and the statistical description of the states of a system are equivalent. But this would mean that the phase trajectory would pass through all points of that surface, which is not true. Nevertheless, the principle question itself (on the measurement of averages) remains. It acquired several formulations united by the term *Boltzmann's ergodic hypothesis* and led to a number of remarkable mathematical theorems which to the present day adorn the theory of dynamical systems.

We recall, for example, the ergodic theorem of George D. Birkhoff (1931).

Let V be an invariant part of phase space with respect to a phase flow that respects the measure μ, and let f be a μ-measurable function on V. If the measure of the part V is finite, then for μ-almost all points of V there exists a time average along the trajectories; it is a μ-measurable function $\tilde{f}$ whose mean value $\tilde{A}$ in V is the same as the mean value A of the function f itself in V.

Furthermore, if V is metrically indecomposable with respect to this phase flow, then the function $\tilde{f}$ is constant μ-almost everywhere (that is, the averages along the trajectories for μ-almost all points $p \in V$ are the same and coincide with the mean value A of the function f on V).

(One says that a space X with a measure μ that is invariant with respect to the flow acting on X is *metrically indecomposable* with respect to the flow and the flow is *ergodic* in X if X cannot be represented as a disjoint union of sets X_1, X_2, each of which is of positive measure and is invariant with respect to the flow.)

It should be pointed out that for the purposes of thermodynamics what are important are not so much the individual ergodic theorems on dynamical systems on fixed spaces of finite dimension, but perhaps the even rougher asymptotic facts on averages relating to passage to the limit (thermodynamic limit) when the number of particles and the corresponding dimension of the phase space increases unboundedly. In this connection it makes sense to analyse more closely the possible interpretations, manifestations and development of the concentration principle considered in Part 2.

3.1.6 *Paradoxes, problems, difficulties*

Usually even a very beautiful theory is a simplified model of a phenomenon that works well only within certain limits which, as a rule, we recognize by going out to the boundaries allowed and running into new problems.

For a statistical theory that reduces thermodynamics to mechanics and predicting, after the second principle of thermodynamics, the evolution of closed systems in the direction of increasing entropy, the first serious touchstone was the question posed by Boltzmann's teacher and now called the *Loschmidt paradox* (1876). It is extremely simply stated and consists in the following.

Any Hamiltonian mechanical system admits time reversal. But how can such a system lead to the selected direction of evolution? In fact, if the direction of time is changed, then the direction of the evolution is changed?!

This question was discussed by better minds at that time (see, for example, the books [1], [4] and the classic survey [6]).

Zermelo added fuel to the flames. (The *Zermelo paradox* 1896. It is due to that same Ernest Zermelo who created a stir among mathematicians and is better known by Zermelo's axiom.) By appealing to Poincaré's recurrence theorem he said that molecules not only can but certainly will (at some moment) all collect in the corner of the room.

And, more generally, any neighbourhood of any point of the phase space of a Hamiltonian mechanical system wandering under the action of the phase flow in a bounded region of the space will return infinitely many times to intersect its original position. (This simple and important observation of Poincaré (1883) is almost obvious if one recalls Liouville's theorem asserting that the phase flow flows like an incompressible fluid, preserving the phase volume. If the images of a neighbourhood over equal intervals of time were pairwise disjoint, then their total volume would be infinite. If, on the other hand, there is an intersection of the mth and nth images and $m < n$, then it is an intersection of the $(n - m)$th image of the neighbourhood with this neighbourhood itself.)

It has to be said that "the mechanical theory of heat lived a difficult life and only achieved its rightful recognition towards the end of the 19th century, primarily thanks to the work of Max Planck in 1887-1892"; see [19], p. 236.

We also note that, as a rule, the classics better understood the merits and defects of their theories, which they themselves most often pointed out. For example, we cite the words of Carathéodory on classical thermodynamics. (Carathéodory, who wrote the article [18a], which incidentally was greatly admired by Planck who stimulated the subsequent publication of the work [18b].)

"One can pose the following question. How should phenomenological thermodynamics be structured so as to use in calculations only directly measurable quantities, that is, volume, pressure and chemical composition of the body ? The theory that arises in this way is logically indisputable and mathematically perfect, since by originating from really observable facts it gets by with the minimum number of hypotheses. On the other hand, it is this very advantage that makes it less useful from the point of view of the researcher, not only because temperature appears in it as a derived quantity, but primarily because the smooth walls that are artificially erected in this theory of structures do not enable one to establish any connection between the world of visible and tangible matter and the world of atoms."

This is an understanding of the need and importance of statistical thermodynamics.

Surprisingly, the first real confirmation of atomism and the molecular structure of matter supposedly turned out to be Brownian motion, discovered by the English botanist Robert Brown as far back as 1828. The mathematical foundations of Brownian motion with its important conclusions for physics were laid down by Einstein (in the year 1905 of his intellectual explosion) by the work [3a] and the subsequent work [3b] , which was already called "The theory of Brownian motion".

In many respects the following opening lines of [3a] are interesting and instructive.

"In this paper it will be shown that, according to the molecular kinetic theory of heat, bodies of microscopic dimensions suspended in a liquid, as a result of the molecular thermal motion must perform movements of such a size that they can be easily detected under a microscope. Possibly these movements are identical with so-called Brownian motion; however, the data available to me with regard to the latter are so imprecise that I cannot form a definite opinion about this.

If this motion along with the expected laws of behaviour is really observed, then classical thermodynamics can no longer be considered to be completely correct for microscopically distinguishable particles and then a precise determination of the true atomic dimensions is possible. If, on the other hand, the prediction of this motion is not justified, then it will be a convincing argument against molecular-kinetic ideas about heat."

([Clausius, Maxwell, Boltzmann and Gibbs had a feeling for the statistical interpretation of the second principle of thermodynamics and defended it. But their explanations were based on thought experiments coming from the postulate of the existence of molecules. Only after the discovery of Brownian motion does the interpretation of the second principle of thermodynamics as an absolute law become impossible. Brownian particles rising and falling as a result of the thermal motion of the molecules is a clear demonstration for us of a perpetual motion machine of the second kind. Therefore at the end of the 19th century the investigation of Brownian motion acquired enormous theoretical significance and attracted the attention of many theoretical physicists including Einstein.] This is what it says in the book [19].)[2]

---

[2] As Bohr said, "not for debate, but for search of the truth": the above citation from the work of Einstein shows that even here not everything is that simple.

Einstein's work on Brownian motion then became the ground of an extensive mathematical theory. Gibbs too did not leave the following generations without a job. Without mentioning the mathematical problems relating to the modern understanding of equilibrium states (as invariant measures) and evolution towards them, we merely recall the *Gibbs paradox* from classical thermodynamics, which led to the inevitability even of the quantum treatment of states.

In the middle of a horizontal cylinder there are two semipermeable movable partitions A and B in that order. The left half of the cylinder is filled with a gas a that can penetrate through the membrane A but not through B. The right-hand part of the cylinder is filled with a gas b that can penetrate through the membrane B but not through A. If the partition A is gradually moved to a stop on the left, then the gas a will stay where it was, while the gas b will spread out throughout the whole cylinder. The entropy of its state is increased, therefore the total entropy of the system of two gases, which is equal to the sum of the entropies of their states, will also increase. We now do the same with partition B, slowly moving it sideways to a stop on the right. The gas a spreads throughout the whole cylinder. The entropy of the system again increases by the same fixed amount ($k\ln 2$).

We now repeat the experiment continually making the characteristics of the gases approach each other. In the end we obtain one gas. Then the initial and final states of the system are indistinguishable. But have their entropies turned out to be different ?

Apparently physical characteristics are not continuous ?! For example, this can be related to energy. We shall try to take this into account.

3.2 Quantum statistical thermodynamics (a few words)

3.2.1 *Calculation of states and the conditional extremum*

Quantum calculations here are even more transparent than classical ones. We shall carry them out following Schrödinger [16].

We consider an ensemble of N identical but numbered systems, each of which is in one of the numbered states $1,2,\ldots,l,\ldots$. Let

$$\varepsilon_1 \leq \varepsilon_2 \leq \ldots \leq \varepsilon_l \leq \ldots$$

be the values of the energy of an individual system in these states, and $a_1, a_2, \ldots, a_l, \ldots$ the number of systems of the ensemble that are in the states $1,2,\ldots,l,\ldots$, respectively.

Such a collection $a_1, a_2, \ldots, a_l, \ldots$ can be realized in many ways. More precisely, the number of ways is equal to

$$P = \frac{N!}{a_1!a_2!\dots a_l!\dots}.$$

The collection of numbers $a_1, a_2, \dots, a_l, \dots$ must satisfy the conditions

$$\sum_l a_l = N, \qquad \sum_l \varepsilon_l a_l = E,$$

where E is the total combined energy of the systems of the ensemble.

We now look for the maximum value of P under the indicated restrictions; this will give us the greatest probability of getting the set of filling numbers $a_1, a_2, \dots, a_l, \dots$.

(We explain that in the cases of interest for thermodynamics when the number N of systems and the number of possible energy levels ε_l is very large we have the concentration phenomenon. It can be shown that the total number of all possible states of the ensemble under our conditions is almost equal to the maximum value of P that we are going to find. Hence, we actually find the maximum probable set of filling numbers $a_1, a_2, \dots, a_l, \dots$.)

By Stirling's formula we have $n! \simeq \sqrt{2\pi n}\left(\frac{n}{e}\right)^n$ for large values of n. Therefore we can assume that $\log(n!) \approx n(\log n - 1)$. (Here log = ln.)

Then, using the method of Lagrange multipliers for finding a conditional extremum, we write

$$\sum_l \log a_l da_l + \lambda \sum_l da_l + \mu \sum_l \varepsilon_l da_l = 0,$$

whence it follows that

$$\log a_l + \lambda + \mu\varepsilon_l = 0 \quad \text{and} \quad a_l = e^{-\lambda - \mu\varepsilon_l}$$

for any l. Here λ and μ are subjected to the conditions

$$\sum_l e^{-\lambda-\mu\varepsilon_l} = N, \qquad \sum_l \varepsilon_l e^{-\lambda-\mu\varepsilon_l} = E.$$

Denoting by $E/N = U$ the average energy required for one system of the ensemble, we can write the result obtained in the following form:

$$\frac{E}{N} = U = \frac{\sum_l \varepsilon_l e^{-\mu\varepsilon_l}}{\sum_l e^{-\mu\varepsilon_l}} = -\frac{\partial}{\partial\mu}\log\sum_l e^{-\mu\varepsilon_l},$$

$$a_l = N\frac{e^{-\mu\varepsilon_l}}{\sum_l e^{-\mu\varepsilon_l}} = -\frac{N}{\mu}\frac{\partial}{\partial\varepsilon_l}\log\sum_l e^{-\mu\varepsilon_l}.$$

Additional considerations explaining the physical meaning of the quantity μ in the thermodynamic situation lead to the fact that

$$\mu = \frac{1}{kT},$$

where k is the Boltzmann constant and T is the absolute temperature. There arises an important quantity, called the *statistical sum*:

$$Z = \sum_l e^{-\frac{\varepsilon_l}{kT}}.$$

We can now write how the filling numbers are distributed in accordance with the energy levels for a given absolute temperature:

$$a_l = N\frac{e^{-\frac{\varepsilon_l}{kT}}}{Z}.$$

The entropy of the system to within an additive constant has the form

$$S = k \log Z + \frac{U}{T} = k \log \sum_l e^{-\frac{\varepsilon_l}{kT}} + \frac{1}{T}\frac{\sum_l \varepsilon_l e^{-\frac{\varepsilon_l}{kT}}}{\sum_l e^{-\frac{\varepsilon_l}{kT}}}.$$

The remaining thermodynamic functions (potentials) also admit similar expressions.

Out of interest we shall calculate how the entropy in the above formula behaves as $T \to 0$. Suppose for generality that corresponding to the lowest energy level are n possible states of the system, that is, $\varepsilon_1 = \varepsilon_2 = \ldots = \varepsilon_n$, and suppose that the following m levels are also the same, that is, $\varepsilon_{n+1} = \varepsilon_{n+2} = \ldots = \varepsilon_{n+m}$.

Then as $T \to +0$ we have

$$S \sim k \log\left(ne^{-\frac{\varepsilon_1}{kT}} + me^{-\frac{\varepsilon_{n+1}}{kT}}\right) + \frac{1}{T}\frac{n\varepsilon_1 e^{-\frac{\varepsilon_1}{kT}} + m\varepsilon_{n+1}e^{-\frac{\varepsilon_{n+1}}{kT}}}{ne^{-\frac{\varepsilon_1}{kT}} + me^{-\frac{\varepsilon_{n+1}}{kT}}},$$

whence it follows that $S \sim k \log n$ as $T \to +0$. In particular, in the physically significant case when $n = 1$ the limit is exactly equal to zero. But even if $n \neq 1$ and n is not very large, the quantity $k \log n$ is almost equal to zero, in view of the smallness of the Boltzmann constant ($k \approx 1{,}38 \cdot 10^{-23}$ J/K).

3.2.2 Clarifying remarks and additional comments

1. First we say a few words about the calculations that we have just made.

The results that we obtained above in a relatively elementary fashion were based on results that are sufficient for finding the conditions for a maximum of the function P, since the overwhelming part of the possible states of the whole ensemble of our systems lies in a small neighbourhood of the maximum, that is, they are the most probable states of the ensemble.

A significantly more thorough detailed analysis of the question was carried out in the work of Darwin and Fowler, which has now become classical. This can be read about in, for example, the sixth chapter of Schrödinger's book [16].

2. We add that, following Gibbs, they introduce the *statistical integral*

$$Z(\beta,a) := \int_\Gamma e^{-\beta H} dq dp,$$

which is analogous to the statistical sum, where, as in the Gibbs distribution, $H(q,p,a)$ is the Hamiltonian and $\beta = 1/k\tau$ with the previous meaning of all the variables.

Then the internal energy of the system and the equations of state of the system (see pp. 85, 101) are expressed in the form

$$E = -\frac{\partial \log Z}{\partial \beta}, \qquad A_i = \frac{1}{\beta}\frac{\partial \log Z}{\partial a_i}, \quad 1 \le i \le n,$$

while its entropy is expressed in the form of the Legendre transform of $\log Z$

$$S = k\left(\log Z - \beta \frac{\partial \log Z}{\partial \beta}\right).$$

The entropy admits another representation, which was already known to Boltzmann:

$$S = -k \int_\Gamma \varrho \log \varrho$$

in terms of the canonical Gibbs distribution. It is in this aspect that it appears in information theory already for an arbitrary distribution density.

Let us verify this formula. Since $\varrho = e^{-\beta H}/Z$, we have

$$-\int_\Gamma \varrho \log \varrho = \log Z + \beta \int_\Gamma H e^{-\beta H}/Z.$$

The last integral is $-\frac{\partial \log Z}{\partial \beta}$. Bearing this in mind and recalling the expression for the entropy given above as the Legendre transform of $\log Z$ we see that our formula is indeed true.

3. We shall demonstrate the remarkable compactness and efficiency of the technique of statistical integrals.

Suppose that in ordinary space $\mathbb{R}^3$ there is a vessel D of volume V filled with an ideal gas consisting of n weakly interacting moving particles of the same mass m. The state of each particle is determined by its position (three coordinates) and its momentum (three more coordinates). The Hamiltonian (kinetic energy) of the system has the form

$$H = \sum_{i=1}^{n} \frac{|p_i|^2}{2m},$$

and the statistical integral

$$Z = \int_{\Gamma} e^{-\beta H} = \int_{D^n} dq \int_{\mathbb{R}^{3n}} e^{-\beta H} dp = V^n \int_{\mathbb{R}^{3n}} e^{-\beta H} dp$$

is easily worked out in spherical coordinates:

$$Z = c\, \frac{V^n m^{3n/2}}{\beta^{3n/2}},$$

where the positive coefficient c depends only on n.

Thus,

$$\log Z = n \log V - \frac{3n}{2} \log \beta + \frac{3n}{2} \log m + \log c.$$

Hence we immediately find the connection between the energy and the absolute temperature of the system

$$E = -\frac{\partial \log Z}{\partial \beta} = \frac{3n}{2\beta} = \frac{3}{2} nk\tau,$$

as well as the pressure

$$P = \frac{1}{\beta} \frac{\partial \log Z}{\partial V} = \frac{nk\tau}{V}$$

as the generalized force corresponding to the parameter V.

If one goes over to the notation of classical thermodynamics, replacing τ by T, k by R/N, nk by MR (where R is the universal gas constant, N is the Avogadro number and M is the quantity of gas in moles), then the relations that one gets become Joule's law $E = \frac{3}{2} MRT$ and Clapeyron's law $PV = MRT$, respectively.

Observe that the original Hamiltonian of the system was not even given to us as a function of the temperature and an external parameter. Furthermore, the calculations do not depend on the number n of molecules. The latter fact may puzzle the physicist who understands that it makes sense to talk about the temperature and pressure of a gas only when n is fairly large.

This combination of merit and strangeness is contained in the axiomatic model of Carathéodory of classical phenomenological thermodynamics that we spoke about earlier. This drew the attention of Planck, who was an admirer of Carathéodory's work.

Was it not a similar situation, along with a lot of other things, that bothered Maxwell, who gave several proofs of his law of the distribution of the molecules according to their speeds (kinetic energies) ? One of the proofs is quite elementary and goes through for any number n of molecules, while the other leads to the result as $n \to \infty$.

4. We now say a few words on the subject matter itself and on these writings.

Almost after each, albeit very small, subsection and section of these writings there usually appears not only the subject, but also a whole area of investigations a detailed account of which is the subject of an independent full-blooded course. We have looked at a great deal, but taking a bird's eye view. One should realize this fact by enriching and refining the knowledge of interest obtained here by using specialist literature. The bibliography given below will provide a certain access to it.

The philosophy of statistical thermodynamics and the deep connections and content of its concepts and principles was the subject in its day of the classical survey work of P. and T. Ehrenfest [6].

We have outlined in modern mathematical language only the skeleton of the two fundamental principles of thermodynamics without touching on the variety of its concrete manifestations and physical thermodynamics itself (see, for example, [20]–[23]).

5. Finally we now add a few words by way of footnotes for which there was no suitable place. We start with the two fundamental concepts of thermodynamics, energy and entropy, and then we shall have finished with them.

It is amusing that, when in 1841 Julius Robert von Mayer attempted to publish his first work on the law of conservation of energy, Peggendorff the publisher of the journal "Annalen der Physik" refused to publish it. Meanwhile, this refusal actually turned out to be a blessing, because in the first edition the article contained so many mistakes that it would have seriously compromised the very idea lying at its foundation; see [19].

Roughly the same thing, but for a different reason, happened with Clausius with his second principle of thermodynamics. He too had to defend the quantity which he later called *entropy*.

(Clausius's choice of the term entropy is explained [19] as follows.

[This quantity is mathematically strictly defined but is not physically intuitive. Furthermore, its absolute value remains undetermined; only its change in thermodynamically isolated irreversible systems is defined; in the ideal case of reversible processes the entropy remains constant.

Since its change is zero for ideal reversible processes and positive for real ones, the entropy is a measure of the deviation of a real process from an ideal one. This explains the name of this quantity given by Clausius, which etymologically means change.]

To tell the truth, this does not seem very convincing to me. For the notation for the difference there exist more precise terms characterizing deviation rather than variation, change and transformation. "En"ergy is in fact the capacity to perform work contained "in" the object, while "en"tropy — is the capacity to perform transformation, mutation, redistribution, mixing contained "in" something. It is not for nothing that the part of the atmosphere where most active phenomena, variations, transformations occur, which we observe every day, is called troposphere.)

6. Some further comments about the two principles of thermodynamics.

The first principle of thermodynamics: this is the general law of conservation of energy and, of course, the original discovery of the equivalence of heat and work, including the mechanical equivalent of heat.

The second principle of classical thermodynamics: this is the assertion on the existence of the function of state entropy and the fact that the evolution of a closed system is directed towards increasing entropy of the system when going from one equilibrium state to another. (A gas contained in half a vessel fills the whole vessel when the partition is removed. It cannot of its own accord move back and collect in the original half of the vessel.)

Under the influence of external actions (for example, when by moving a piston we change the volume of a gas in a cylinder) the system (gas) already can change its state in another direction. If everything happens infinitesimally slowly, then we can assume that all the time we pass through equilibrium states of the system and the entire process is reversible.

Mathematically this is expressed in the fact that for reversible processes the integral of the form $T^{-1}\delta Q$ or $\tau^{-1}\omega$ of the reduced heat along the path of transfer is equivalent to the increase of entropy of the system.

If, on the other hand, the equilibrium condition is violated in the course of the process, then the increase of entropy of the system turns out to be larger. This inequality (of Clausius) characterizes the property of irreversible processes (for example, the contact heat exchange of bodies at different temperatures).

For any realizable thermodynamic cycle (closed path) γ Clausius's inequality has the form

$$\int_{\gamma} \frac{\delta Q}{T} \geq 0;$$

here equality holds only on reversible transitions γ. They go through equilibrium states of the thermodynamic system, which we considered earlier (see p. 86).

We observe that formally entropy is defined only for equilibrium states of a dynamical system. The extension of this concept to non-equilibrium states is usually accompanied by a division of the system into microsystems which, by assumption, are in an equilibrium state, and then summing the entropies of these systems. We also emphasize that in the formulation of the second principle of thermodynamics (either in the form of Clausius on the contact passage of heat from a hot body to a cold one, or in the form of Carathéodory on the adiabatic unattainability of certain states) it is tacitly assumed that one is dealing with thermally homogeneous systems, that is, systems such that thermal equilibrium is possible in them only under equal temperatures of all parts of the system.

A violation of this condition can lead to apparent contradictions and paradoxes. Take, for example, a pair of identical bars. One is at a temperature of 0 degrees and the other at a temperature of 100 degrees. If they are brought into thermal contact, then equilibrium occurs when both have a temperature of 50 degrees. One does not obtain a transfer of more heat from the second body to the first. But if one divides each of the bars in half by a thermally impenetrable barrier, then by slight manipulations connected with successive shifts and contacts of the bars one can bring the situation to a state when the halves of the first bar will have temperatures of 75 and 50 degrees, respectively, whereas the halves of the originally hot bar will have temperatures of 50 and 25 degrees. After this the barriers can be removed. The temperatures of the bars become constant, and the first bar is hotter than the second, which was the source of the heat. An apparent violation of Clausius's principle!

We give another example, again relating to Clausius's principle. Two ideal gases with different heat capacities taken in amounts of 1 mole each are separated by an adiabatic sliding piston. You can verify (following T.A. Afanas'eva-Ehrenfest) that for this thermally inhomogeneous system (parts of which can have different temperatures in the equilibrium state) the form δQ does not have an integrating factor, therefore there is no hindrance to adiabatic passages between any states of the system ([21], p. 59).

7. Certain dates (apart from Big Bang, birth of the Sun, and Prometheus's heroic deed).

Phenomenological thermodynamics

1680–1705 The invention of the steam engine. The first patent for the steam engine was issued to the English metal worker Thomas Newcomen.

1765 The question of James Watt (1736–1819) on the coefficient of useful action of a heat engine (subsequently the mechanical equivalent of heat).

1824 The work of Sadi Carnot (1796–1832) "On the driving force of fire" (subsequently the second principle of thermodynamics).

1834 The discovery by Benoit Clapeyron (1799–1864) of the work of S. Carnot and Clapeyron's law (combining the laws of Gay-Lussac and Boyle–Marriott).

1840–1850 Statement and refinement of the law of conservation of energy: Julius Robert Mayer (1814–1878), James Prescott Joule (1818–1889), Hermann von Helmholtz (1821–1894), Rudolf Clausius (1832–1888).

1850–1925 The setting up of classical phenomenological thermodynamics. Equilibrium thermodynamics as a science (Clausius). The contact geometry of thermodynamics (Gibbs). The axiomatization of thermodynamics (C. Carathéodory).

Molecular theory of heat

(The idea of the molecular structure of matter is ancient. The idea of explaining heat as molecular motion is more recent: ... Francis Bacon (1561–1626), Johannes Kepler (1572–1630), Leonard Euler (1707–1783) ...)

1738 Daniel Bernoulli (1700–1782). Molecular explanation of pressure (in his work "Hydrodynamics").

1856 August Krönig (1822–1879). Connections between temperature and kinetic energy.

1857–1865 Rudolf Clausius (1832–1888). The developed theory of heat. Analysis of the principle of the equivalence of heat and work (1850). The concept of internal energy; the formulation of the law of conservation in the form $\delta Q = dE + PdV$; the concept of entropy and the second principle of thermodynamics (1865).

Statistical mechanics, thermodynamics, physics

1860–1866 James Clerk Maxwell (1831–1879). Maxwell's law of the distribution of molecules according to their speed and kinetic energy; the mean free path of molecules; Maxwell's demon, questions of reversibility.

1868–1906 Ludwig Boltzmann (1844–1906). Boltzmann's distribution of molecules according to their energy in a potential field. Consistent statistical approach to thermodynamics. Boltzmann's equation and the evolution of thermodynamic systems. Entropy and the probability of a state. The H-theorem on increase of entropy. Problems of the second principle of thermodynamics. The ergodic hypothesis.

1883–1892 Henri Poincaré (1854–1912). Dynamical systems. The recurrence theorem. A gas as a collisionless continuous medium and its evolution.

1896 Ernest Zermelo (1871–1953). Paradoxes of thermodynamics.

1902 John Willard Gibbs (1839–1903). Mathematical theory of statistical mechanics. Measures, distributions and their evolution in Hamiltonian systems. Equilibrium states as invariant measures.

1905 Albert Einstein (1879–1955). The theory of Brownian motion and the second principle of thermodynamics; sizes of atoms, the Avogadro number.

Quantum statistical mechanics, thermodynamics, physics.

1887–1892 Max Planck (1858–1947) The birth of quantum theory.

1902 John Willard Gibbs (1839–1903) The thermodynamic paradox.

1926 Erwin Schrödinger (1887-1961). Quantum statistical mechanics.

References Part 1, Chapters 1 & 2[3]

Available Textbooks

[1] P.W. Bridgeman, *Dimensional analysis*, Yale Univ. Press, New Haven CT, 1932.
[2] Garrett Birkhoff, *Hydrodynamics. A study in logic, fact and similitude* (Revised edition), Princeton University Press, 1960.
[3] L.I. Sedov, *Methods of similarity and dimensional methods in mechanics* (6th ed.), Nauka, Moscow, 1967.
[4] M.H. Holms *Introduction to the Foundation of Applied Mathematics*, Springer-Verlag, Berlin, 2009.

Reference books

[4a] *International system of units (SI)*, Vysshaya Shkola. Moscow, 1964.
[4b] *Units of quantities*: Dictionary-Handbook, Standards Publishing House, Moscow, 1990.
[4c] *Le Système International d'Unitées (SI)*, Edition du Bureau International des Poids et Mesures, Paris, 1970.

History of the Problem

[5] H. Görtler, "Zur Geschichte des Π-Theorems", ZAMM **55**:1 (1975), 3–8.
[6] A. Federman, "Some general methods of integrating first-order partial differential equations ", S. Peterburg Politekh. Inst. Otdel. Estestvozn. i Mat., **16**:1 (1911), 97–155.
[7] From the letters to the editors: M. Rozhkov, "N.A. Morozov as a founder of dimensional analysis", Uspekhi Fiz. Nauk **49**:1 (1953), 180–181.

Supplementary literature

[8] D.D. Landau and I.M. Lifschitz, *Course in theoretical physics vol. 6. Fluid mechanics* (2nd ed.), Pergamon Press, Oxford, 1987.
[9] N.E. Kochin, I.A. Kibel and N.B. Roze, *Theoretical Hydrodynamics, Part* 2, Fizmatlit., Moscow, 1963.
[10] H. Whitney, *Collected Papers, Volume II*, (James Eells, Domingo Toledo, eds.), Birkhauser, Boston, 1992, 530–584.
Reprint from: "The mathematics of physical quantities. Part I, Mathematical models for measurement", Am. Math. Monthly **75** (1968), 115–138; "Part II, Quantity structures and dimension analysis", ibid, 237–256.
[11] V.I. Arnold, *Mathematical methods of classical mechanics*, Nauka, Moscow, 1989. (p.50, Similarity arguments); English transl., Graduate Texts in Math., Vol. 60, Springer-Verlag, Beriln-New York, 1978.
[12] M. Jarman, *Examples in quantititive zoology*, Edward Arnold (Publishers) Ltd., London 1970.
[13] J. M. Smith, *Mathematical ideas in biology*, Cambridge University Press, Cambridge, 1968.

[3] Each chapter has its own numbering and bibliography (apart from the auxiliary Chapter 2 of Part 1). The list and its marking are conditional and far from complete or perfect, not to say arbitrary, if one takes into account the sea of literature on the questions touched upon here. (They say that on the entrance to one of the bars in Texas the following notice is hung: "Please don't shoot at our musicians. They are doing the best they can.")

[14] Yu.I. Manin, *Mathematics and Physics*, Znanie, Moscow, 1979. (What is new in Science and Technology, Series Mathematics and Cybernetics, vol.12.) See also the book Yu.I. Manin, *Mathematics as metaphor*, MCCME, Moscow, 2008; English transl.,*Selected Essays*. American Mat. Soc., Providence, RI, 2007.
[15] L.V. Ovsyannikov, *Group analysis of differential equations*, Nauka, Moscow, 1978.

Part 1, Chapter 3

[1] A.N. Kolmogorov, "Local structure of turbulence in an incompressible viscous liquid with very high Reynolds numbers", Dokl. Akad. Nauk SSSR **30**:4 (1941), 299–303. Also published in Uspekhi Fiz. Nauk **93**, 1967, 476–481;
and in the book A.N. Kolmogorov *Collected Works. Mathematics and Mechanics* Nauka, Moscow, 1985, 281–287.
(See Kolmogorov 100th Birthday Issue Vol. 1, Bibliography, Fizmatlit, Moscow, 2003. There is also a German translation in the collection. *Statistiche Theorie der Turbulenz*, Akademie-Verlag, Berlin, 1958, 71–76.
[2] A.N. Kolmogorov, "On the degeneracy of isotropic turbulence in an incompressible viscous liquid, Akad. Nauk SSSR, **31**:6 (1941), 538–541.
Also published in the book: A.N. Kolmogorov *Collected Works. Mathematics and Mechanics*, Nauka, Moscow, 1985, 287–290.
There is also a German translation in the collection. *Statistiche Theorie der Turbulenz*, Akademie-Verlag, Berlin, 1958,147–150.
[3] A.N. Kolmogorov, "Dissipation of energy under local isotropic turbulence", Dokl. Akad. Nauk, **32** (1941), 19–21;
and in the book A.N. Kolmogorov *Collected Works. Mathematics and Mechanics* Nauka, Moscow, 1985, 290–293.
[4] A.N. Kolmogorov, A refinement of previous hypotheses concerning the local structure of turbulence in a viscous incompressible fluid at high Reynolds number", J. Fluid Mech. **13**:1, 1962.
[5] A.N. Kolmogorov, "Mathematical models of turbulent motion of a viscous liquid", Uspekhi Mat. Nauk **59**:1, 2004, 5–10.
(Posthumous publication. The whole of this issue is devoted to Kolmogorov. In it are published the texts of a number of lectures given at the grand international conference "Kolmogorov and Modern Mathematics" in Moscow in 2003 and timed for the 100th birthday of A.N. Kolmogorov. On pp. 25–44 there is the text of V.I. Arnold's lecture "Kolmogorov and Natural Science", where, in particular, Kolmogorov's general hydrodynamical principles, which initiated the study of turbulence, are explicitly stated.)
[6] A.M. Obukhov, "On the distribution of energy in the spectrum of a turbulent flow", Izv. Akad. Nauk SSSR, Seriya geogr. i geofiz. 1941, 4–5.
[7] A.M. Obukhov, "Some specific features of atmospheric turbulence", J. Fluid Mech. **13**:1 (1962) 77–81; J. Geophys. Res., **67** (1962) 3011–3014.
[8] A.M. Obukhov, "Kolmogorov flow and laboratory simulation of it", Uspekhi Mat. Nauk, **38**:4 (1983) 101–111.
[9] *Studies in turbulence*, Nauka, Moscow, 1994.
(Collection of articles by authors participating in the work of the seminar of A.M. Obukhov and A.S. Monin. Devoted to the memory of A.M. Obukhov.
Preface by Academician O.M. Belotserkovskiĭ with reminiscences on Professor S.G. Kalashnikov who gave lectures on Physics at Moscow State University.)
[10] D.D. Landau and I.M. Lifschitz, *Course in theoretical physics vol. 6. Fluid mechanics* (2nd ed.), Pergamon Press, Oxford, 1987.
[11] N.E. Kochin, I.A. Kibel' and N.B. Roze, *Theoretical Hydrodynamics, Part* 2, Fizmatlit., Moscow, 1963.
[12] *Hydrodynamic instability*, Collection of articles, Mir, Moscow, 1964.
(Papers by E. Hopf, Garrett Birkhoff, J. Kampé de Feriét, R. Kraichnan et al; and the bibliography in those papers.)

[13] M.A. Lavrent'ev and B.V. Shabat, *Problems of hydrodynamics and their mathematical models*, Nauka, Moscow, 1977.

[14] *Visions in Mathematics* Towards 2000. Part 1.
GAFA (Geom. funct. anal.) Special volume – GAFA2000, Birkhauser–Verlag, Basel, 2000.

[15] A. Kupiainen, "Lessons for turbulence", GAFA, Geom. funct. anal., Special volume – GAFA2000, 316–333.

[16] U. Frish, *Turbulence. The legacy of A.N. Kolmogorov*, Cambridge University Press, Cambridge, 1995.

[17] Kolmogorov, his 100th anniversary, vol. 1, Fizmatlit, Moscow, 2003. Bibliography, 225, 70–80, 342.

[18] D. Ruelle, *Hasard et chaos*, Edition Odile Jacob, 1991.

[19] S. Mallat, *A wavelet tour of signal processing*, Second ed., Academic Press, 1999.
(See p.236 on the work of Kolmogorov on turbulence in the years 1941 and 1962.
It is noted that the model so far does not take into account the turbulent process and the alternation of turbulent and relatively calm regimes of flow. There are no explanations of the mechanism of the formation of vortices and the interchange of energies between small-scale and large-scale structures.
"The understanding of the properties of hydrodynamic turbulence is the most important problem of modern physics.
No single formula is in a position to construct a physico-statistical description based on the Navier–Stokes equations, which might have given us the possibility of understanding the global phenomenon of turbulent flows, as has been done in thermodynamics".)

[20] Quotation from the newspaper "Izvestiya" on 15 April 2005, p.17:
[Academician Alekseĭ Lipanov, director of the Institute of Applied Mathematics said in an interview with a correspondent of "Izvestiya":
"We have succeeded in creating a mathematical model of turbulence, thus solving a problem considered over the centuries to be insoluble, since the process of turbulence was considered to be random. We have succeeded in proving and showing that this is a fallacy and that this process lends itself to accurate modelling. The practical application of this discovery is obvious; constructors of aeroplanes and ships will now be able use this technique for preventing catastrophes ..."
From the Kaliningrad International conference on selected problems of modern mathematics timed to coincide with Karl Jacob's 200th birthday.]
We hope that contact between the authors of items [19] and [20] will take place and will not lead to turbulence.
Incidentally, has possibly an important practical question of interest to many hydrodynamicists, physicists and engineers at the same time been solved ? This is the question of the overfall of pressure at the ends of a tube taking into account the possible turbulent drag of the flow of a liquid or gas being driven through the tube.
(In this connection see, for example, the paper by Barenblatt in the same issue of Uspekhi Mat. Nauk as item [5]. In connection with turbulence in general it may be of interest to look at the themes of the lectures at the recent international symposium Hamiltonian Dynamics, Vortex Structures, Turbulence, held at the Steklov Institute, Moscow in August 2006, as well as the themes of the lectures at the international conference "Mathematical Hydrodynamics" held at the Steklov Institute, Moscow in June 2006.)

It is also helpful for mathematicians to look at the following items.

[21] Albert Shiryaev, "On the classical, statistical, and stochastic approaches to hydrodynamics and turbulence", Thiele centre for applied mathematics in natural science, Research report, 02 I January 2007.

[22] V.I. Yudovich, "Global solubility against collapse in the dynamics of an incompressible fluid" Mathematical events of the 20th century, Fazis, Moscow, 2003, 519–548.

[23] S. Friedlander and V. Yudovich, "Instabilities in Fluid Motion", Notices of the AMS, **46**:11 (1999), 1358–1367.
[24] V.I. Arnold and V.A. Khesin, "Topological methods in hydrodynamics", Applied Mathematical Sciences, **125**, Springer–Verlag, Berlin, 1998.

Part 2, Chapter 1

[1] V.A. Kotel'nikov, On the transmission capacity of the "ether" and the wire in electrocommunication. All-union energy committee. All-union congress on questions of the reconstruction of communications and the development of the low-current industry, Izdat. Upravleniya svazi RKKA, Moscow, 1933.
[2] C.E. Shannon, "A mathematical theory of communication", Bell System Tech. J. **27** (1948), 379–423, 623–656.
[3] *Information theory and its applications*, Fizmatlit, Moscow, 1959.
This is a collection of translations and includes the articles:
R.V.L. Hartley, "Transmission of information", BSTJ **7**:3, 535–563, 1928.
B.M. Oliver, J.R. Pierce and C.E. Shannon, "The philosophy of P.C.M. Proc. IRE **36**:11, 1324–1331, 1948.
W.G. Tuller, "Theoretical limitations on the rate of transmission of information", PIRE **37**:5 (1949), 468–478.
C.E. Shannon, "Communication in the presence of noise", PIRE **37**:1, 10–21, 1949.
Y.W. Lee, T.P. Cheatham and J.B. Weisner, "Application of correlation analysis to the detection of periodic signals in noise", PIRE **38**, 1165, 1950.

Survey with development and bibliography.

[4] Ya.I. Khurgin and V.P. Yakovlev, *Functions with compact support in Physics and Technology*, Nauka, Moscow, 1971.

Part 2, Chapter 2

Mathematical primary sources.

[1] H. Poincaré, *Calcul des probabilités*, Gauthier Villars, Paris, 1912.
[2] P. Levy, *Problèmes concrets d'analyse fonctionnelle*, Gauthier Villars, Paris, 1951.
[3a] V. Milman and G. Schechtman, *Asymptotic Theory of Finite Dimensional Normed Spaces.* (With an Appendix by Gromov.) Lecture Notes in Mathematics, **1200**, Springer-Verlag, 1986.
[3b] V.D. Milman, "Phenomena that occur in high dimensions", Uspekhi Mat. Nauk **59**:1, 157–168, 2004; English transl. in Russian Math. Surveys **59**:1 (2004), 159–169.
[3c] E. Milman, "On the role of convexity in isoperimetry, spectral gap and concentration", Invent. Math., **177**, 1–43, 2009.
[3d] K. Ball, *An Elementary Introduction to Modern Convex Geometry. Flavors of Geometry.* MSRI Publications, Volume **31**, 1997.

Physical sources and their mathematical development.

[4] H.A. Lorentz, *Thermodynamics*, Lectures on Theoretical Physics Vol 2, MacMillan, London, 1927.
[5] E. Schrödinger, *Statistical thermodynamics*, Cambridge University Press, Cambridge, 1946.
[6] A.Ya. Khinchin, "Symmetric functions on many-dimensional surfaces", *Collection of articles to the memory of A.A. Andronov*, Izdat. Akad. Nauk SSSR, Moscow, 1955, 541–56.
[7] V.I. Opoitsev, "Non-linear law of large numbers", Avtom. Telemekh. **1994**:4, 65–75; English transl. in Autom. Remote Control **55**:4, 511–519.

[8] Yu.I. Manin, *Matematika i Fizika*, Znanie, Moscow, 1979. (Novie v nauke i tekhnike, Ser. Mat., Kibernetika, vol. 12)
It is also in the book
Yu.I. Manin, *Mathematics as metaphor*, MCCME, Moscow, 2008; English transl.,*Selected Essays*. American Mat. Soc., Providence, RI, 2007.

[9] R.A. Minlos, *Introduction to mathematical statistical physics*, Amer. Math. Soc. Providence, R.I., 2000.

[10] M. Kac, *Probability and related questions in physical sciences*, Summer Seminar in Applied Mathematics, June 23 – July19, 1957, Boulder, CO; Lectures in Applied Mathematics , Vol. 1.A, Amer. Math. Soc., Providence, RI, 1976.

[11] V.V. Kozlov, *Thermal equilibrium in the sense of Gibbs and Poincaré*, Izdat. RKhD, Moscow-Izhevsk, 2002.

[12] D. Ruelle, *Chance and chaos*,Princeton University Press, Princeton, NJ, 1991.

[13] J. Kurchan and L. Laloux, "Phase space geometry and slow dynamics", J. Phys. A: Math. Gen. **29** (1996) 1929–1948. Printed in the UK.

Part 2, Chapter 3

Primary Sources.

[1] C.E. Shannon, "A mathematical theory of communication", Bell System Tech. J. **27** (1948), 379–423, 623–656.

[2] *Information theory and its applications*, Fizmatlit, Moscow, 1959.
Collection of translations including the article
C.E. Shannon, "Communication in the presence of noise", PIRE **37**:1 (1949), 10–21.

Diversion on ϵ-entropy and Hilbert's problem.

[3] D. Hilbert, *Mathematische Probleme. Gesammelte Abhandlungen*, **III**, 1935, 290–329.
See also H. Weyl, *David Hilbert and his mathematical work*, Bull. Amer Math. Soc. **50** (1944), 112–154.

[4] A.G. Vitushkin, "On Hilbert's thirteenth problem", Dokl. Akad. Nauk SSSR **95**:4 (1954), 701–704.
See also A.G. Vitushkin, "Hilbert's thirteenth problem and related questions", Uspekhi Mat. Nauk **59**:1 (2004), 11–24; English transl. in Russian Math. Surveys **59**:1, (2004), 11–25.

[5] A.N. Kolmogorov, "Estimates of the minimal number of points of ϵ-nets in various function classes and their application to the problem of representing functions of several variables as superpositions of functions of a smaller number of variables", Dokl. Akad. Nauk SSSR **10**:2 (1955), 192–194.

[6a] A.N. Kolmogorov, "On the representation of continuous functions of several variables as superpositions of functions of a smaller number of variables", Dokl. Akad. Nauk SSSR **108**:2 (1956), 179–182; English transl. in Amer. Math. Soc. Transl. (2) **17** (1961), 369–373.

[6b] A.N. Komogorov, "On the representation of continuous functions of several variables in the form of a superposition of continuous functions of one variable and addition", Dokl. Akad. Nauk SSSR **114**:5 (1957), 953–956; English transl. in Amer. Math. Soc. Transl. (2) **28** (1963), 55–59.

[7a] V.I. Arnold, "From Hilbert Superposition Problem to Dynamical Systems", Proceedings of 1977 Conferences at Fields Institute, Fields Institute Communications **24**, (1999), 1–18.
The Russian text of this report is in the book *Mathematical events of the twentieth century*, Fazis, Moscow, 2003, 19–48.

[7b] V.I. Arnold, "On functions of three variables", Dokl. Akad. Nauk SSSR **114**:4 (1957), 679–681.

[8] A.N. Kolmogorov and V.M. Tikhomirov, "ϵ-entropy and ϵ-capacity of sets in function spaces", Uspekhi Mat. Nauk **14**:2 (1959), 3–86; English translation in Amer.Math. Soc. Transl. II. Ser. 17 (1961), 227 – 364.

Supplementary literature.

[9] A.G. Vitushkin, *Estimate of the complexity of the problem of tabulating*, Fizmatlit, Moscow, 1959.

[10] V.I. Buslaev and A.G. Vitushkin, "Estimate of the length of code of signals with a finite spectrum in connection with sound-recording problems ", Izv. Akad. Nauk SSSR, Ser. Mat. **38** (1974), 867–895; English transl. in Math. USSR-Izv. **8** (1974), 867–894.

[11] K. Blatter, *Wavelet analysis. A primer*, A.K. Peters, Natick, MA, 1998.

[12] S. Malla, *Wavelets in signal processing*, Mir, Moscow, 2005.

[13] J.R. Higgins, "Five short stories about the cardinal series", Bulletin of the Amer. Math. Soc. (New Series) **12** (1985), 45–89.

[14] Ya.I. Khurgin and V.I. Yakovlev, *Functions with compact support in physics and technology*, Nauka, Moscow, 1971.

[15] E.S. Wentsel, *Probability theory*, Nauka, Moscow, 1964; English transl., Mir, Moscow, 1982.

[16] A.S. Kholevo, *Introduction to quantum information theory*, MCCME, Moscow, 2002.

Part 3, Chapter 1

Primary sources, translations and descriptions.

[1] *Second principle of thermodynamics*, Gostekhizdat, Moscow-Leningrad, 1934. (A collection of works by S. Carnot, W. Thomson-Kelvin, R. Clausius, L. Boltzmann and M. Smolukhovskiĭ.)

[2] H.A. Lorentz, *Thermodynamics*, Lectures on Theoretical Physics Vol 2, MacMillan, London, 1927.

[3] P. Ehrenfest and T. Ehrenfest, "Begriffische Grundlagen der Statistischen Auffassung in der Mechanik", Enzyclopädie der Math. Wiss. Bd. IV, 1911.

[4] M. Gliozzi, *Storia della fisica*, Storia delle Scienze Vol. 2, Turin. 1965.

[5] M. Planck, "Forlesungen über Thermodynamik", **5**, Aufl. Leipzig: Veit, 1917.

[6] A. Sommerfeld, *Thermodynamik und Statistik*, Vorlesung über Teoretische Physik, Band V, Wiesbaden, 1952.

[7] M. Leontovich, *Introduction to thermodynamics. Statistical physics*, Nauka, Moscow, 1983.

[8a] The Feynman lectures on physics, Volume 1 (Richard P. Feynman, Robert. B. Leighton and Matthew Sands), Addison—Wesley Publishing Company, INC., Reading, MA, Palo Alto, London, 1963

[8b] R. Feynman, *Statistical mechanics: a set of lectures (notes taken by R. Kiguchi and H.A. Feiveson, edited by Jacob Shaham)*, Benjamin, Reading, MA, 1972.

[9] E. Schrödinger, *Statistical thermodynamics*, Cambridge University Press, Cambridge, 1946.

[10] E. Fermi, *Thermodynamics*, Dover Publications, New York, 1937.

[11a] C. Carathéodory, "Untersuchungen über die Grundlagen der Thermodynamik", Mathematische Annalen **67**, (1909), 355–386.
See also the Russian translation *Razvitie sovremennoi fiziki*, Nauka, Moscow, 1964, where there is also the article:
M. Born, "Critical comments on traditional accounts of thermodynamics", ibid, 223–257.
[Included in: C. Carathéodory,*Gesammelte Mathematische Schriften*, Munich, 1955, Vol.2, 131–166.

[11b] C. Carathéodory, "Über die Bestimmung der Energie und der absolutin Temperature mit Hilfe von reversiblen Prozessen. Sitzungs-berichte der Preussischen Akademie der Wissenschaften Physikalisch-mathematische Klasse, Berlin, 1925, 39–47.

Part 3, Chapter 2

Thermodynamics and contact geometry

[1a] J.W. Gibbs," Graphical methods in the thermodynamics of fluids", Transactions of Connecticut Academy, **1873**:2, 309–342.
[1b] J.W. Gibbs, "A method of geometrical representation of the thermodynamic properties of substances by means of surfaces", Transactions of Connecticut Academy, **1873**.
[1c] J.W. Gibbs, "On the equilibrium of heterogeneous substances", Transactions of Connecticut Academy, **1876, 1878**.
[1d] J.W. Gibbs, "Elementary principles in statistical mechanics: developed with especial reference to the rational foundation of thermodynamics", New Haven, CT, 1902.
[2] Proceedings of the Gibbs Symposium held at Yale University, Newhaven, CT, May 15–17, 1989, American Mathematical Society,Providence, RI: American Institute of Physics, New York, 1990.
[3] V.I. Arnold, "Contact geometry: the geometrical method of Gibbs's thermodynamics", Proceedings of the Gibbs Symposium held at Yale University, New Haven, CT, May 15–17, 1989, American Mathematical Society, Providence, RI; American Institute of Physics, New York, 1990, 163–179.

Mathematical aspects

[4] H. Cartan, *Differential calculus. Differential forms*, Hermann, Paris 1967.
[5] P.K. Rashevskiĭ, "On a criterion for any two points of a totally non-holonomic space to be joined by an admissible line", Uch. Zap. Ped. Inst. Libknekht, Ser. Fiz. Mat. Nauk **2**(1938), 83–94.
[6] W. Chow, "Sisteme von linearen partiallen Differentialgleichungen erster Ordnung", Math. Ann. **117**, 1939, 98–105.
[7] "Sub-Riemannian geometry", Progr. Math. **144**, Birkhäuser, Basel, 1996.
[8] H.J. Sussmann, "Orbits of families of vector fields and integrabilities of distributions", Trans. Amer. Math. Soc. **180**, 1973, 171–188.
[9] M. Gromov, "Carnot-Carathéodory spaces seen from within. Sub-Riemannian geometry", Progr. Math. **144**, Birkhäuser, Basel, 1966, 79–323.
[10] R. Montgomery, "A tour of subRiemannian geometry, their geodesics and applications. Mathematical Surveys and Monographs", Amer. Math. Soc. vol. **91**, 2002.
[11] F. Warner, *Foundations of differentiable manifolds and Lie groups*, Scott Forseman, Glenview, IL,1971
[12] V.I. Arnold, *Mathematical methods of classical mechanics*, Nauka, Moscow, 1989; English transl., Graduate Texts in Math., Vol. 60, Springer-Verlag, Berlin-New York, 1978.
[13] V.I. Arnold and A.V. Givental, *Symplectic geometry*, Izdat RKhD, Izhevsk, 2000.
[14] H. Lieb and J. Yngvason, "The mathematics of the second law of thermodynamics", Geom. Funct. Anal., Special volume GAFA, 2000, 334–358.
[15] A.B. Antonevich, V.I. Bakhtin , A.V. Lebedev and V.D. Sarazhinskiĭ, "Legendre analysis, thermodynamic formalism and spectra of Peron-Frobenius operators", Dokl. Akad. Nauk **390**:3 (2003), 295–297; English transl. in Dokl. Math. **67**:3 (2003), 343–345.

Part 3, Chapter 3

Some works of the classics

[1] L. Boltzmann, *Wissenschaftliche Abchandlungen*, I–III, Chelsea Publishing Company, New York, NY, 1968. (Reprint of a work first published 1909 in Leipzig.)

[2] J.W. Gibbs, "Elementary principles in statistical mechanics: developed with especial reference to the rational foundation of thermodynamics", New Haven, CT, 1902.

[3a] A. Einstein, "Über die von molekularkinetischen Theorie der Wärme geforgerte Bewegung von in rühenden Flüssigkeiten suspendierten Teilchen", Ann. Phys. **17** (1905), 549–560.

[3b] A. Einstein, "Zur Theorie der Brownschen Bewegung", Ann. Phys. **19** (1906), 371–381.

[4] H. Poincaré, *Thermodynamique.* Deuxieme èdition, revue et corrigeé, Gauthier-Villar, 1908.

[5] H.A. Lorentz, *Thermodynamics*, Lectures on Theoretical Physics Vol 2, MacMillan, London, 1927.

[6] P. Ehrenfest and T. Ehrenfest, "Begriffische Grundlagen der Statistischen Auffassung in der Mechanik", Enzyclopädie der Math. Wiss. Bd. IV, 1911.

Further sources

[7] N.N. Bogoliubov, *Problems of dynamical theory in statistical physics*, Gostekhiizdat, Moscow-Leningrad, 1946.

[8] A. Ya. Khinchin, *Mathematical foundations of statistical mechanics*, Izdat. RKhD, Izhevsk, 2002.

[9] F.A. Berezin, *Lectures on statistical physics*, MCCME, Moscow, 2007.

[10] V.P. Maslov, *Ultra-secondary quantization and quantum thermodynamics*, Editorial URSS, Moscow, 2000.

[11] M. Kac, *Probability and related questions in physical sciences*, Summer Seminar in Applied Mathematics, June 23 – July19, 1957, Boulder, CO; Lectures in Applied Mathematics , Vol. 1.A, Amer. Math. Soc., Providence, RI, 1976; includes Appendix I from the lectures of G.E. Uhlenbeck "The Boltzmann equation", 2nd ed., Izdat. URSS, Moscow, 2003.

[12] V.V. Kozlov, *Thermal equilibrium in the sense of Gibbs and Poincaré*, Izdat. RKhD, Moscow-Izhevsk, 2002.

[13] R.A. Minlos, *Introduction to mathematical statistical physics*, Amer. Math. Soc. Providence, R.I., 2000.

[14a] Ya.G. Sinaĭ, *Modern problems of ergodic theory*, Fizmatlit, Moscow, 1995.

[14b] Ya.G. Sinaĭ, *Theory of phase transitions*, Izdat. RKhD, Moscow-Izhevsk, 2001.

[15a] D. Ruelle, *Chance and chaos*,Princeton University Press, Princeton, NJ, 1991.

[15b] D. Ruelle, *Thermodynamic formalism. Mathematical structures of classical equilibrium statistical mechanics*, Addison-Wesley, Reading MA, 1978.

[16] E. Schrödinger, *Statistical thermodynamics*, Cambridge University Press, Cambridge, 1946.

[17] A. Sommerfeld, *Thermodynamik und Statistik*, Vorlesung über Teoretische Physik, Band V, Wiesbaden, 1952.

[18a] C. Carathéodory, "Untersuchungen über die Grundlagen der Thermodynamik", Mathematische Annalen **67**, (1909), 355–386.
See also the Russian translation *Razvitie sovremennoi fiziki*, Nauka, Moscow, 1964, where there is also the article:
M. Born, "Critical comments on traditional accounts of thermodynamics", ibid, 223–257.
[Included in: C. Carathéodory, *Gesammelte Mathematische Schriften*, Munich, 1955, Vol.2, 131–166.

[18b] C. Carathéodory, "Über die Bestimmung der Energie und der absolutin Temperature mit Hilfe von reversiblen Prozessen. Sitzungs-berichte der Preussischen Akademie der Wissenschaften Physikalisch-mathematische Klasse, Berlin, 1925, 39–47.

[19] M. Gliozzi, *Storia della fisica*, Storia delle Scienze Vol. 2, Turin. 1965.

[20] The Feynman lectures on physics, Volume 1 (Richard P. Feynman, Robert. B. Leighton and Matthew Sands), Addison—Wesley Publishing Company, INC., Reading, MA, Palo Alto, London, 1963
[21] M.L. Leontovich, *Introduction to thermodynamics. Statistical physics*, Nauka, Moscow, 1983.
[22] E. Fermi, *Thermodynamics*, Dover Publications, New York, 1937.
[23] L.D. Landau and E.M. Lifschitz, *Course of theoretical physics, Vol. 5: Statistical physics*, Pergamon Press, London-New York-Paris, 1958.

Appendix

Mathematics as language and method

(Why mathematics ? Explanation mainly for non-mathematicians)[1]

1. As legend has it (the popularization of which we owe, according to some sources, to the remarkable Frenchman Arouet, better known under the pseudonym Voltaire[2]) an apple fell on the head of Newton. Since at that time his head also contained Kepler's laws and much else, the result proved to be different from what is obtained when the apple falls on other heads.

Some of what has happened in mathematics as a science after this event has, over the past three centuries, become the alphabet of natural science. When looked at from above the labyrinth always seems simple.

2. With a view to saying here something about mathematics in general, I begin with an example which will make the subsequent mutual understanding easier.

Older than the Newton apple legend is that of Archimedes running naked through Syracuse shouting "Eureka, eureka!" This legend has some variants. I shall give two of them and then make some observations and conclusions.

I quote Academician M.L. Gasparov.[3]

"Here is how it was. The Syracuse tyrant Hieron obtained from a goldsmith a golden crown and he wanted to check whether the goldsmith was not mixing silver into the gold. It was necessary to compare the volumes of the crown and a piece of pure gold of the same weight. Archimedes, by sinking into a bath flooding to the edges and seeing how the water displaced by his body overflowed the edges, suddenly realized that it is easy to measure the volumes of two bodies of different shapes."

A somewhat more advanced version is the following.[4]

The goldsmith was given gold and silver separately which were alloyed together to form a solid object (crown). Archimedes is required to find out whether or not the goldsmith had replaced part of the gold with silver.

Let x_1, x_2 be the amounts of gold and silver in the ready-made crown. Its entire weight $x_1 + x_2 = A$ was easy to measure and verify that it was equal to the total weight given to the goldsmith. The goldsmith is no fool

---

[1] In the novel "The Master and Margarita " by Mikhail Afanasievich Bulgakov the magic perfomance ends with a disclosure, as is well known. The popular article "Mathematics as language and method" has a similar relation to "Mathematical Analysis of Problems in the Natural Sciences".

[2] The surname Arouet with the extension L(e)J(eune) (that is, Jr.) looks in Latin letters like AROVETLI, from which VOLTAIRE comes as an anagram.

[3] "Entertaining Greece", Fortuna Limited, 2002, p.362.

[4] Possibly this latter version was less appropriate for the lucid, highly professional, fascinating book by Mikhail Leonovich Gasparov (1935–2005) of revered memory, but is more appropriate for us now.

and this equality is found to be true. We now suspend the crown on a spring balance and immerse it in a full bath. Collecting the overflowed water we measure its volume V and weight P; we also read off the new reading (on the spring balance) of the weight B of the immersed crown. We now tie all this data together. Some readers who find it too difficult can skip a couple of sentences.

The densities p_1, p_2 (masses per unit volume in terms of the weight on Earth) of our precious metals have been known for a long time. Then the quantities $x_1/p_1 = V_1$, $x_2/p_2 = V_2$ give the volumes of each metal in the crown. Hence, $V_1 + V_2 = V$.

Thus we already have the relations

$$\begin{cases} x_1 + x_2 = A, \\ x_1/p_1 + x_2/p_2 = V. \end{cases}$$

If mathematics has taught us not only to write but also to solve a system of equations of the form

$$\begin{cases} a_{11}x_1 + a_{12}x_2 = b_1, \\ a_{21}x_1 + a_{22}x_2 = b_2, \end{cases}$$

then we find our unknowns, fulfil the state orders, obtain a reward for continuing our livelihood and, what is probably most important, get such a delight from our discovery that is out of this world, that we dash through Syracuse repeating "Eureka! Eureka!"

Scientists are, as a rule, freedom-loving people, but are ready to give up their lives for relating something to something.

Not being too lazy to work out the remaining data of the experiment we find that $A - B = P$, that is, a body immersed in water loses the same amount of weight as the weight of the water displaced by it.

But this is, in fact, ARCHIMEDES'S LAW!

(Naturally, all this can be repeated with another liquid or even with a gas.)

This is worth more than the crown of the Syracuse tyrant Hieron! Thanks to him for the problem, a by-product of which overshadows the question.

Yes, rafts, boats and ships sailed even before the days of Archimedes. But now we can design a ship and foretell its freight capacity up to the time it is lowered into the water. Now we can also design a dirigible for transporting by air large constructions (in the construction of rigs and observatories) in places that are inaccessible for surface transport, and so on.

3. We have given this example so that, with that as a background, we can say a few words about the specific character of mathematics as a science. Here we shall not attempt to give a definition of mathematics. We simply observe the example and ascertain something lying already on the surface.

Mathematics has enabled us to translate a question into some special language (some kind of symbols, equations, ...). Hence mathematics has the attribute of a *language*.

But here there is a noticeable difference from the translation of the original question, for example, from Greek to Russian or Chinese. In each such translation, on the one hand the original text and content of the original question can somehow or other be recovered, while on the other hand, in such a translation only the way of writing is changed; the question remains.

By going over to a mathematical text, that is, to mathematical notation, we absolutely lose the ability to return to the specific part of the question if we have lost its original text. But then we obtain a certain mathematical question (here it is the solution of a system of equations) which, once solved, answers both our original particular question as well as similar questions at one go.

Mathematics finds methods of solving systems of equations and many other problems which at first glance are of interest to nobody apart from mathematicians themselves. But in actual fact, like a number, it services a wide sphere of concrete objects and phenomena.

Thus, mathematics mostly gives not only a special *language* (in which one endeavours to write the question arising, by throwing out everything of secondary importance), but also a *method* of solving the problem, which is now a purely mathematical one. In particular, by solving it we also obtain the answer to the special question of interest to us.

Now we are prepared to mention and evaluate the following statements:

A great book of nature is written in the language of mathematics. (Galileo)

If the notation is suitable for discoveries, then the path to the truth is strikingly shortened. (Leibniz)

Mathematics is the art of calling different things by the same name. (Poincaré)

We give one further quotation from Gasparov's book, which we recalled earlier:

"Placed on the grave of Archimedes in accordance with his will, in place of an epitaph there is a drawing of a cylinder with a ball inscribed in it and written down is the ratio of their volumes 3:2 discovered by him. One hundred and fifty years later, when the famous Roman writer Cicero served in Sicily, he still saw this memorial, forgotten and overgrown with blackthorn".

There, where the Greek Archimedes had earlier lived, had long ago no longer been Greece. Not only the graves of the great disappeared, but also entire countries and civilizations. But Archimedes's law lives on along with nature and the universe. In this unity is the everlasting value and fascination of true knowledge and science.

In mathematics, of course, one can perceive many other aspects. For example, Lomonosov noted (not without justfication) that "mathematics puts the mind in order". It also teaches one to hear an argument and respect the truth.

4. Mathematics, for all its apparent abstractness, feeds on problems of natural science and generously gives back fruits raised in this fertile soil. This is breathing in and out. Disturbing this balance is dangerous both in science and in teaching. Science desiccated by scholasticism perishes.

5. And now something else in this connection.

A typical mathematical statement looks like this:

Theorem. *If such and such, then such and such.*

In another notation this is: $A \Rightarrow B$ (A implies B). Usually textbooks on mathematics very carefully inspect the fragment ($\Rightarrow B$), that is, they present a detailed logical irreproachable proof of the fact that B does indeed follow from A.

Only very naive and inexperienced people, including some mathematicians, can allow themselves to be glad at such a theorem if the original premise A is empty, uninteresting or unnatural. In any case it must be borne in mind that A is absolutely on an equal footing and is a very essential informal element of the mathematical theorem.

We explain this by recalling one of the rare sentences which, according to the testament of eye-witnesses, was uttered by the taciturn Gibbs (creator of the mathematical foundations of classical thermodynamics and statistical mechanics) at the time of a discussion at Yale University, where he worked, of the role and place of mathematics, its axiomatic method, and also its interaction with physics and natural science.

Gibbs stood up and said: "A mathematician can allow himself to assume anything he likes, whereas a physicist is not allowed to abandon common sense."

Gibbs sat down. And this concluded his part of the discussion.

Of course Gibbs did not express without irony that which the outstanding mathematician and thinker Hermann Weyl wrote 50 years later:

"The constructions of the mathematical mind are at the same time free and necessary. The individual mathematician is free to define his concepts and establish his axioms as he pleases. But the question is whether he will interest his colleagues-mathematicians by the products of his imagination. We cannot but feel that some mathematical structures, thanks to the joint efforts of many scholars, carry the imprint of necessity which is not affected by the chances of their historical appearance. Everyone who contemplates the spectacle of modern algebra will be struck by the way freedom and necessity complement each other in that discipline".

The absorption by mathematics of new problems (inhalation), the subsequent interpretation, solution, generalization and, on this basis, the construction of a new abstract mathematical theory (exhalation) is the natural closed cycle of the working of this science. In one historical period the process of accumulating factual material dominates, and in another period it is putting together the results and distributing everything according to the positions,[5] or logical ordering and formalization.[6] Furthermore, this can sometimes be observed both in individual areas and even in the creative work of one and the same prominent mathematician (who in one period can do and advocate one thing and in another period, something else; this is even not hypocrisy, provided, of course, that the fact itself is not disclaimed).

6. A landmark advance in science is often achieved (or, better still, takes shape) by the interesting and distinctive way it especially brightly emerges in such of its abstract areas as theoretical physics and mathematics.

Imagine an hour glass. In order for it to work it has to be turned "from foot to head" from time to time. In mathematics it is the same. First one obtains several new interesting facts. Among them is revealed something that in some or other respect is central and key, connecting a lot of earlier material. This is taken as an original principle turning everything from foot to head (for example, by making a theorem an axiom) and development continues by relying now on this new principle, heralding a large area of facts of mathematics and the universe.

For example, Newton's laws grew out of the discoveries of Galileo and Kepler. On the other hand, by taking Newton's laws as the basis, we can obtain from them Kepler's laws and much else. The subsequent development of physics led to the new variational principles of mechanics, including the large area of phenomena and interactions distinct from the interactions described by central forces.

In a certain sense, at such moments of "turning over the hour glass" a change of scale occurs. Here it consists precisely in the change and reduction of the number of basic principles and, at the same time, a broadening of the field of objects and phenomena that they embrace and unify.

(Incidentally, those who do not keep an eye on such an opportune change of scale usually talk about overloading the school curriculum.

One more thing, possibly out of place. There are two ways to become rich: to lay ones hands on property —-war, robbery; to create valuables— daily conscientious work.

There where they often talk about grandeur they prefer to use the first method.

The outstanding Dutch physicist and teacher of many other physicists Lorentz modestly said about the first world war (as testified by Einstein[7]): "I am fortunate that I belong to a nation that is too small to carry out big acts of stupidity".

---

[5] For example, at the turn of the 19th century the Encyclopaedia of Mathematical Sciences published in Germany under the initiative of F. Klein.

[6] Even later are the multi-volume works of N. Bourbaki.

[7] A. Einstein, Collection of scientific works, Vol. 4, Nauka, Moscow, 1967, the article "G.A. Lorentz as creator and human being", 334–336.

They say that relatively recently (and possibly even now) written in the primers of Japanese children is roughly the following: "Our country is small and poor. We need much and must work well in order to become richer".

Finland, the former backward province of Russia, demonstrates to us the effectiveness of honest work and respect for the law. The inductive method of science is somewhat more solid and reliable than the principle "Everything, now, and for free".)

7. We now say a few words about the so-called higher mathematics.

So what is characteristic for this higher mathematics, which took shape, roughly speaking, in the works of Newton and Leibnitz and which has been energetically developed over the past three centuries since then?

This mathematics has been taught to deal with not only constant quantities, but also with *processes in evolution.*

In general, what has arisen and gradually taken shape is the concept of functional dependence or *function,* which has proved fundamental for science as a whole.

The rate of change of any variable quantity and acceleration have obtained an adequate mathematical formulation in terms of *derivatives.*

There arose a new language and a new *calculus* (*differential* when, with respect to a given dependence, one seeks the rate of change of a quantity, and *integral* when one is solving the inverse problem: when from a record of the speed or acceleration one seeks the location of a moving object, as though it were a submarine.)

The fundamentals of this differential and integral calculus are, like the multiplication table, essential components of any scientific education, or indeed, education in general.

Let us explain why. If $x = x(t)$ is the law of motion, that is, the dependence of the coordinates of the object on time, then differential calculus enables one to find its velocity $v = x'(t)$ and acceleration $a = x''(t)$.

If the mass m of the object and the force F acting on it are known, then by Newton's law the following equation must hold:

$$mx'' = F,$$

which is an equality involving derivatives (in the present case, the second derivative $x''(t)$ of the original function $x(t)$). If we are interested in how this body will move under the action of the given force F, then we shall look for the unknown dependence $x = x(t)$ satisfying the above equation.

Thus, one is required to investigate and solve an equation of a completely new type: a *differential equation.*

Again this is already a purely mathematical question (abstracted from the motions of planets, evolution of star systems, the working of nuclear reactors, baking bread, growth of bank savings and microorganisms, insurance situations, populations of fish and animals, predators and prey; and so on and so forth...); but it is a question that relates directly to all of this.

Thus, when mathematics constructs a theory by suggesting methods for solving a certain class of problems it calls on us for tools servicing in one fell swoop a whole new sphere of concrete phenomena. These phenomena can possibly be from an earlier time, such as rafts before Archimedes law. But now we understand them better. More precisely, we constructed a mathematical model for them which we understand and which, with the help of the mathematical tools, we know how to control. Already this alone has, as a rule, applications such that they repay with interest the lavish outlay by civilized society on chalk for mathematicians.

As Hermann Weyl remarked: "In all natural sciences knowledge is based on observation. But observation only ascertains the position of things. How does one foresee the future ? For this, observation has to be combined with mathematics".

Without mathematics, of course, there would be no Newton, no Maxwell, no Einstein, no Bohr,..., as we know them. Hence there would be no civilization whose fruits we gladly and easily use daily. To make this quite clear, imagine for the moment that we humans lost the gift of speech, words and language.[8] I am not arguing whether this is a good thing or a bad thing. I simply wanted to explain an alternative and the place of mathematics.

[8] Possibly for some people a more striking loss would be the loss of a mobile or a T.V. or other such objects; as Socrates said, "how many things there are that we can do without".

Unforeseen epilogue

The English translation of this book was almost completed when Vladimir Igorevich Arnold suddenly passed away. I do not know who will be reading this book, but the book and all of us mathematicians missed the most bright, thorough and professional reader, to whom many places of the text were addressed explicitly or implicitly and with a friendly smile.

A lot will yet be told and written about V.I. Arnold. Objective time and the next generations will estimate his merits. But the next generations are not given the chance to be his contemporaries. They will not be able to realize that the loss of Arnold for mathematics, at least in Russia and especially with the current state of science in the country, is similar to the loss of habitat.

Printing: Ten Brink, Meppel, The Netherlands
Binding: Stürtz, Würzburg, Germany